On Reflection 1 *– moored yacht in the creek in Old Leigh just after sunrise*
at low tide with mirrored reflection
• 70mm lens, f22.0, 1/15 sec, ISO 100, WB6000

Mike Drave

Weir at Stoneham's Lock – *a view from behind the weir next to the lock on the Chelmer and Blackwater canal*
• *18-50mm lens at 38mm, f22.0, 0.6 sec, ISO 100, WB 6200*

Waterscape Essex

A Pictorial Journey of
The Coast and Inland Waterways of Essex

Michael Drane

charles publishing

For my dear sons
Jonathan and Daniel

ISBN 978-0-9572110-0-1

Published by:
Charles Publishing
66 Hallowell Down
South Woodham Ferrers
Chelmsford, Essex CM3 5GZ
Tel: 01245 329021
Email: mikedrane@talktalk.net

FRONT COVER
Sea Wall – *stretching for miles along this part of the Essex coast
from Bradwell round to Burnham-on-Crouch*
• *70-300mm lens at 70mm, f16, 1/50 sec, ISO 100, WB 5150,
ND Graduated Filter*

BACK COVER
Blackwater Estuary – *at low tide from Heybridge Basin early one
morning just after sunrise with the mist clearing in the distance*
• *18-50mm lens at 50mm, f16.0, 1/40 sec, ISO 100, WB 6200,
ND Graduated Filter*

Baddow Mill Lock 1 – *view of the lock taken early one spring
morning on the Chelmer and Blackwater canal*
• *55-200mm lens at 88mm, f22.0, 1/8 sec, ISO 100, WB 6200*

Contents

Thames Barge Race – *'Reminder' and 'Decima' competing on the Blackwater estuary*
• 55-200mm lens at 85mm, f9.0, 1/320 sec, ISO 100, WB 5150

Preface

Having been born, bred and living for over sixty years in Essex I have yet to explore the entirety of the county. Fate has delayed this for my latter years for my enjoyment as a photographer to try and record for the future generations of like minded people.

To record every inch for a single volume in a short space of time would be an impossible task for one person, but hey, I am going to give it my best shot (excuse the pun). Maybe a sequel or two would be a better solution - God willing!

This book contains a broad cross section of the county concentrating mainly on its delightful fretted coastline, estuaries and including some of the gently rolling countryside. It is a pictorial extravagance keeping the narrative to a minimum (apart from the historical inclusions) so that you can cruise through the views and soak up the ambience of the county.

As John Constable, speaking of the Essex countryside, once said 'I love every stile and stump . . . as long as I am able to hold a brush I shall never cease to paint them.'

Drawing from his inspiration my passion is for photographic images which portray a rich variety of 'scape, wildlife and atmospherics. For some of my images I have combined my artistic flare with the creative features of digital photography to produce pleasing fine art images. Thereby, I have blended photography into my other pastimes, which obviously includes walking, as a further addition to my enjoyment of life.

And, for you budding photographers I have included one or two useful tips.

Every view in this book I can offer for sale as prints and are hand printed by myself to capture the spirit or the emotion of the subject using the highest quality materials and techniques. For further details see my website at:

www.photoartanddecor.co.uk
email: mikedrane@talktalk.net

Clacton Pier – one moody afternoon at almost high tide
• *18-50mm lens at 46mm, f16.0, 1/13 sec, ISO 100, WB 5150, ND Graduated Filter*

Ancient History of Essex

... most civilised kingdom in the British Isles with the only city in Britain established and thriving more than 2000 years ago ...

The Fingringhoe Wick Romano-British Port – *This now extinct small river port on the Colne estuary served the legionary fortress and later Roman colony at Camulodunum (Colchester). On the evidence of Roman military equipment and coinage recovered from gravel workings on the Fingringhoe headlands, it is thought that this port was operational very early during the Roman occupation of the Islands, and enabled direct communication with the seaports near the mouth of the Rhenus (Rhine) on the other side of the Oceanus Britannicus (English Channel), where it is known that contemporary storage depots and other Roman naval installations existed.*
• 18-55mm lens at 46mm, f22.0, 1/60 sec, ISO 100, WB 5150

Source: Ashingdon Parish Council
www.essexinfo.net/ashingdonparish/history

The area which Essex now occupies was ruled by the Celtic 'Trinovantes' tribe and its centre of power was the large established city of Caer-Colun (now called Colchester) which was said to be the only city in Britain.

Emperor Julius Caesar came here in 55 BC to meet the then king of the only established centre of power and civilisation in Britain.

Nearly one hundred years later in 43 AD Emperor Claudius also came to Colchester to meet the then king of the Trinovantes and to exercise his rule over what he considered to be their client king and in turn to rule Britain. Consequently Caer-Colun was Romanised to Camulodunum.

The Dark Ages: *after the Romans left in about 420 AD, the Saxons came to this part of Britain and occupied and ruled this region as a new kingdom and renamed Camulodunum to Colun-ceaster or Colne-ceaster. Hence the names of the River Colne and Colchester.*

Various Saxon groups established seven kingdoms in England which survived over several hundred years from about 500 AD until about 1000 AD.

Those 7 Saxon Kingdoms were called the Heptarchia. They were: Essex, East Anglia, Kent, Sussex, Wessex, Mercia and Northumbria.

The Kingdom of Essex (or Kingdom of the East Saxons) consisted of present day Essex, all of London, Middlesex, most of Hertfordshire and parts of present day Suffolk and Cambridgeshire. The rest of Cambridgeshire was either in East Anglia, under the sea, or in Mercia. For a while Kent belonged to The Kingdom.

The Kingdom of Essex is known to have been in existence from 527 AD to 825 AD when Mercia and then Wessex took over the Kingdom. During that 300 years period, there were at least 18 Kings of Essex, from King Aescwine, the earliest known king in 527 AD until King Sigered in 825 AD. Two of those 18 Essex Kings became saints. They were Saint Sigeberht and Saint Sebbi.

Essex now has largely the same boundaries as it had in pre-Roman times, when it was the Kingdom of the Trinovantes. It was the most civilised kingdom in the British Isles with the only city in Britain established and thriving more than 2000 years ago.

Famous rulers included Cunobelin (the 'Cymbeline' of Shakespeare), King Coel (Old King Cole) and Arturo (Arthur) whom the Romans dealt with.

Is there any connection between Colchester and King Arthur?

There is a strong claim based on historic, geographic and place names for the Camelot of King Arthur to be Camulodunum.

The reasons being:

- There was a Trinovantian ruler called Arturo. Trinovantes was the most civilised and established region and Camulodunum was said to be the only city in Britain.

- The Romans came to Camulodunum to visit the ruler and to negotiate with and ultimately dominate Trinovantes and in turn colonised all of Britain.

- The name Camelot is very close to the abbreviated form of the Roman and pre-Roman name Camulod, or its grammatical form Camulodunum.

- There is no other place in Britain called Camelot or similar, nor any place name derived from it, other than Colchester's previous name.

- Historians have no true idea where Camelot is in its often assumed location in the West Country.

- The proximity of the English Channel and 'foreign lands'.

Whereas some argue that Arthur was a semi-legendary character who led the resistance against the invading Anglo-Saxons. Therefore any historical Arthur could only have lived in the west or north of Britain as Colchester lay in the heart of enemy territory.

Colchester Castle – *undeniably one of the most important historic buildings in the country. Colchester was the first capital of Roman Britain and beneath the Castle are the remains of the most famous Roman buildings, the Temple of Claudius.*
Today if you lay your hand on the stonework of the temple it can be said that you are touching the very foundation of Roman Britain.
To Romans the temple was a symbol of their power and success, but to the native Britons it was a symbol of oppression. The temple became a main target of the rebels led by Queen Boudica who attacked the Roman town of Colchester in AD 60. The town's citizens barricaded themselves into the temple but after two days they were all killed.
It is estimated that up to 30,000 people could have been killed during the sacking of Colchester. After the revolt was suppressed the town and its magnificent temple were rebuilt.
Around 1076 William I ordered a royal fortress to be built at Colchester. The great stone base of the ruined Roman temple was an obvious foundation for the central tower, or keep, of the castle.
The huge size of the temple meant that the keep of Colchester Castle was the largest ever built in Britain and is the largest surviving example in Europe.

For most of its life the Castle was used as a prison. One of the most infamous episodes in its history occured in 1645 when Matthew Hopkins, the self-styled Witchfinder General, used the Castle to imprison and interrogate suspected witches.
The Castle first opened to the public as a museum in 1860. Today it is still a living vibrant place.
It is not only the town's flagship museum, but it is also in a real sense a symbol of Colchester, Britain's oldest recorded town.
Source: Colchester Museums
www.colchestermuseums.org.uk/castle/castle_history

- *10-20mm lens at 11mm, f16.0, 1/60 sec, ISO 100, WB 5150, ND Graduated Filter*

Dedham Vale

'I associate my careless boyhood with all that lies on the banks of the Stour. Those scenes made me a painter and I am grateful.'

John Constable

What a superb location to start with – the **Dedham Vale and Stour Valley** which embraces one of our most cherished landscapes.

Picturesque villages, rolling farmland, rivers, meadows, ancient woodlands and a wide variety of local wildlife combine to create what many describe as the traditional English lowland landscape.

Villages in the area include East Bergholt, Capel St. Mary, Stratford St. Mary, Bentley, Dedham, Langham, Lawford and Flatford.

River Stour – a winters view meandering through Dedham Vale
• 18-50mm lens at 46mm, f22.0, 1/30 sec, ISO 100, WB 6700, ND Graduated Filter

This designated Area of Outstanding Natural Beauty stretches upstream from Manningtree to within one mile of Bures.

The landscape quality of the remainder of the Stour Valley has resulted in its designation as a *potential* AONB or Special Landscape Area.

Because much of East Anglia's traditional grasslands have already been drained and ploughed for arable farming, the hedgerows and wildflower meadows of Dedham Vale are among some of England's most precious and vulnerable pastoral landscapes.

The countryside is enhanced by narrow lanes and characteristic timber-framed and thatched houses.

Dedham Vale

Dedham Church – A view across rural farmland towards St Mary the Virgin, home to an original Constable painting 'The Ascension'
- *18-50mm lens at 50mm, f22.0, 1/15 sec, ISO 100, WB 5300, ND Graduated Filter*

LEADING THE EYE THROUGH THE COMPOSITION

Lead-in lines draw the viewer into the scene and also make the image appear more dynamic. Lead-in lines can be found in most landscapes, so look out for hedges, walls, paths, or even patterns in sand or earth.

These lines work best if they lead towards the main focal point, or at least a point of interest.

See how the boarding in the foreground points your eye towards the leaning tree and, once there, how the riverbank takes over and leads the eye to the top and out of the picture.

River Bank – A view heading towards Flatford Mill on the Essex side of the River Stour
- *18-50mm lens at 29mm, f22.0, 1/30 sec, ISO 100, WB 6500*

Dedham Vale

Set on the Essex/Suffolk border the area was made famous by the paintings of England's 18th-century foremost landscape artist, John Constable. He painted many idyllic views of the area in his famous six-foot canvases, scenes which remain easily recognisable today.

The little riverside hamlet of Flatford is the setting for some of Constable's most famous paintings, such as 'The Hay Wain', 'The Mill Stream', 'Boat-building near Flatford' and 'The White Horse'.

From his home in East Bergholt, a mile or so to the north, a young Constable used to walk across the riverside meadows to Dedham every day on his way to school. Possibly passing by Bridge Cottage, adjoining Flatford Bridge, which is now home to a small exhibition on Constable with a NT tearoom and shop.

Flatford Mill, a Grade I listed 18th century watermill built in 1733 by Abram Constable, and Willy Lott's house are owned by the NT but leased to the Field Studies Council which runs arts-based courses there.

Attached to the mill is a 17th century miller's cottage which is also Grade I listed.

The Mill was left to John Constable's father when Abram died.

Flatford Mill – the title of one of Constable's most iconic paintings viewed from the Essex side of the river in sepia tone
• *18-55mm lens at 28mm, f22.0, 1/20 sec, ISO 100, WB 5150, ND Graduated Filter*

Water Power – A close up view of the power behind the mill
• *150-500mm lens at 150mm, f22.0, 1/10 sec, ISO 100, WB 6700*

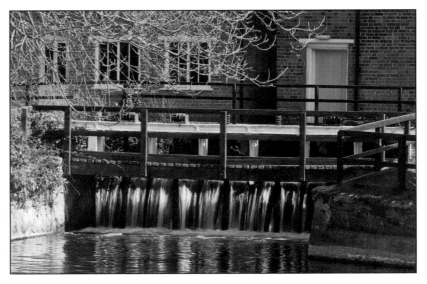

Bridge Cottage *– at Flatford Bridge viewed from the Essex side of the river*
- *18-55mm lens at 49mm, f16.0, 1/30 sec, ISO 100, WB 5150*

Approach to Flatford Bridge *– the River Stour as it meanders through Dedham Vale approaching the bridge*
- *18-50mm lens at 35mm, f22.0, 1/25 sec, ISO 100, WB 7000, ND Graduated Filter*

The Stour Estuary

. . . one of the most important estuaries in Britain for wintering birds . . .

The *River Stour* after passing through Dedham Vale and Constable Country reaches its estuary at Manningtree and Mistley through to Wrabness and Parkeston.

It then converges with the *River Orwell* estuary before flowing out to the North Sea at Harwich on the Essex side and Felixstowe docks on the Suffolk side.

Bridge View – *looking back up the river towards Constable Country from the road bridge near Manningtree where the river becomes the estuary* • *70-300mm lens at 92mm, f16, 1/60 sec, ISO 100, WB 5150*

Mistley Walls 1 – *view towards Manningtree at low tide, the entire eight miles of the Stour estuary are a magnet for artists, walkers, birdwatchers, cyclists, sailing enthusiasts and of course photographers* • *18-50mm lens at 18mm, f16, 1/50 sec, ISO 100, WB 5150, ND Graduated Filter*

Manningtree (Englands smallest market town) together with the neighbouring village of **Mistley** are often viewed as little more than the gateway to Constable country. That may be so but both have their own rich history and architectural heritage, making them well worth taking a little time to explore.

Described in the mid-19th Century as 'an improving market town', Manningtree still hosts a market in the town square twice a week and has its own museum.

Mistley was once the site of a proposed spa and though the project ultimately foundered, the village retains a number of related features icluding the Swan Fountain and Mistley Towers.

For two centuries, Manningtree and Mistley were important brewing centres and ports. Even today, although on a reduced scale, the brewing industry is still well represented.

Mistley Walls 3 – *all along the shoreline are moored small boats of all shapes and sizes*
• *18-50mm lens at 28mm, f16, 1/40 sec, ISO 100, WB 5150*

Witchfinder General

One of Manningtree's more dubious claims to fame is that it had been the home and burial place of the famous Witchfinder General, Matthew Hopkins, who is believed to have been responsible for the killing of around 300 women between the years 1644 and 1646. Many of these women were accused of witchcraft by children.

He carried out his interrogations mainly at inns in Manningtree and Mistley as well as Colchester Castle, with trials being held in Chelmsford.

East Anglia was known as the 'Witch Country'.

Hopkins is believed to have died and been buried in Mistley in 1647. His memory inspired at least one film, Witchfinder General (1968).

The Swan Colony 1 – *once the second largest herd of swans in the country their numbers have now dropped to 250*
• *18-50mm lens at 29mm, f16, 1/60 sec, ISO 100, WB 5150, ND Graduated Filter*

Black Swan – *The Colony are mainly Mute swans which are white but there is occasionally an odd black swan which come from down river. The colony has declined over the years possibly because the nearby Maltings have closed down, meaning the swans cannot live off the grain which was deposited into the river. Swan Rescue, a volunteer organisation now takes care of the swans*
• 70-300mm lens at 300mm, f9, 1/400 sec, ISO 200, WB 5150

The entire eight miles of the estuary is a magnet for artists, walkers, birdwatchers, cyclists, sailing enthusiasts and of course photographers.

From the waterfront there are fantastic views across the estuary surrounded by delightful scenery and you can also see the famous Swan Colony.

It also forms part of the 81 mile Essex Way, a long-distance path stretching diagonally across the county from Epping to Harwich.

The Swan Colony 2 – *attracting hordes of birdwatchers along The Walls throughout the year. The RSPB runs regular events here for the 'twitchers'*
• 18-50mm lens at 50mm, f16, 1/50 sec, ISO 100, WB 5150, ND Graduated Filter

The Stour Estuary

One of the most famous contributors to the area was Richard Rigby (son of Richard Rigby Esquire) who had lavish plans to turn Mistley into a spa town. He engaged the famous architect Robert Adam, but his plans were never brought to fruition as the money ran out.

Mistley Towers and the Swan Fountain are perhaps the most notable remains of Adam's work.

St. Mary's New Church – built around 1870 in New Road and viewed from Mistley Towers (site of the old church)
- *70-300mm lens at 70mm, f16, 1/50 sec, ISO 100, WB 5150*

Mistley Towers – the twin towers of the now demolished Church of St. Mary the Virgin.

The original Georgian parish church on the site had been built in classical style early in the 18th century following the death of Richard Rigby Esquire.

Later in that century his son, the wealthy politician Richard Rigby, planned to transform Mistley into a spa town.

Rigby wished to see a church from the windows of his mansion and a suitably grand church was required for the affluent visitors expected to patronise the new spa.

Thus in 1776, the great architect Robert Adam was commissioned to enhance the church. His design was in the neoclassical style, with a tower at both the east and the west ends of the church.

These are now all that remain of the once magnificent structure.

The design of the towers creates the impression that the building was once more of a miniature cathedral than a parish church.

The main body of the church was small and occupied the (now empty) site between the two towers. It was a single story structure with a simple hipped roof and an entrance portico at its centre. This part of Adam's church was demolished in 1870, when the new parish church in New Road was built.

The remaining towers are Grade I listed
- *18-50mm lens at 21mm, f16, 1/60 sec, ISO 100, WB 5150*

Port of Mistley – offers both ship owners and clients a competitive, efficient and cost effective alternative for UK imports and exports.

This small port provides the ideal location for short sea and transhipment cargoes, with excellent links to Europe, the Baltic, Scandinavia as well as to London, the Midlands and the North.

The port handles a diverse range of cargoes including bulk products (grain, fertiliser and agri-products, aggregates, industrial minerals, and recyclables), forest products, granite, steel products, metals, and various bagged, palletised and unitised cargoes
• 18-50mm lens at 38mm, f9, 1/1600 sec, ISO 200, WB 5150

Felixstowe Docks – viewed from Wrabness, a small village located six miles west of Harwich.

On the right is the start of Wrabness Nature Reserve (52 acres) and Copperas Bay.

The site was once a former mine depot established in 1921 by the Ministry of Defence. It was closed in 1963.

Following closure, a number of planning applications were put forward (including an application for a prison in 1968 and 1989).

The site was saved from closure when it was bought by Wrabness Nature Reserve Charitable Trust in 1992. The site has now been taken over by the Essex Wildlife Trust.

The reserve is an important wildlife site - owls, yellowhammers, whitethroats, turtle dove, song thrush, nightingales and bullfinches can be seen. There are also many wild plants such as corn mint, hairy buttercup, sea aster and ox-eye daisy
• 70-300mm lens at 141mm, f16, 1/125 sec, ISO 100, WB 5150, ND Graduated Filter

The Stour Estuary

Wrabness Nature Reserve at Copperas Bay lies immediately west of the port facilities of Parkeston and is managed by the Royal Society for the Protection of Birds (RSPB).

Consisting of two divergent habitat types: intertidal mudflats (fringed by saltmarsh and estuarine reeds), and 130 acres of deciduous woodland, mainly oak and coppiced sweet chestnut.

At this point it is one of the most important estuaries in Britain for wintering birds, with internationally important numbers of grey plovers, knots, redshanks and dunlin, but most significantly for the numbers of Black Tailed Godwit.

Brent Geese – *can be seen along all of the estuary. It is an Amber List species because of the important numbers found at just a few sites*
• *70-300mm lens at 300mm, f9, 1/500 sec, ISO 200, WB 5150*

Copperas Bay 1 – *Vast areas of intertidal mud and saltmarsh make up Copperas Bay at Wrabness. The bay in winter offers the spectacle of masses of wading birds and wildfowl visiting the Stour Estuary*
• *18-50mm lens at 18mm, f22.0, 0.4 secs, ISO 100, WB 4800, ND Graduated Filter*

Copperas Bay 2 – *Viewed in the other direction towards the mouth of the estuary with the woodland running down to the waterside*
• *18-50mm lens at 18mm, f22.0, 0.4 secs, ISO 100, WB 4800, ND Graduated Filter*

Primrose – *(Primula vulgaris). Early spring flowering woodland plant in Stour Wood, Wrabness. Visitors in mid-summer may see the White Admiral butterfly in the only area in Essex where this species is known to occur*
• *18-50mm lens at 50mm, f2.8, 1/1250 sec, ISO 100, WB 4900*

Stour Wood and Copperas Wood are both extensive areas of sweet chestnut coppice, intermixed with hornbeam and other tree species indicating an ancient woodland area.

Unusually, for Essex, Stour Wood runs right down to the waterside.

In the spring, nightingales and other birds fill the woods with their songs. Spring flowers are also particularly beautiful here icluding a spectacular show of wood anemones in March/April.

*Ramsey Windmill – a Grade II listed post mill, with a three storey roundhouse, built originally in Woodbridge, Suffolk and relocated to the village of Ramsey in 1842.
The village stands on a creek of the River Stour and commands a fine view over the river*
• 18-50mm lens at 20mm, f22.0, 1/20 secs, ISO 100, WB 5150

Essex Windmills

Windmilling in Britain is thought to have begun in the late twelfth century, by which time water mills were already well established. By 1825-35 when windmilling in Essex was at its peak, there were about 285 mills in the county. One hundred years later, only a handful were still at work and by 1950 the last working mill had stopped.

The decline of both wind and water mills from about 1850 followed the arrival of the steam-driven roller mill and improvements in sea, rail and road transport. Grain could be brought from abroad to the huge dockside mills and the new roller-milled white flour could be distributed easily, even to remote country areas.

Windmill types – three types of windmill were commonly used in Britain and all can be found in Essex:

THE POST MILL

With the post mill, the sails are built into the wooden body which houses the machinery. The whole mill body is pivoted on a massive wooden post, allowing the body and hence the sails to be turned to face the wind.

THE TOWER MILL

The main structure of the tower mill is built of brick or stone and so cannot be rotated. The sails are mounted in a separate wooden cap which is arranged so that it can turn on the top of the tower.

THE SMOCK MILL

The smock mill works in the same way as the tower mill, using a rotating cap, but differs in that its main structure is built of timber and is usually octagonal, on a masonry base.

Parkeston Quay – *viewed from Old Harwich Quayside*
• *18-50mm lens at 50mm, f11, 1/500 sec,*
ISO 200, WB 5150

Parkeston Quay

A sea port situated about one mile up-river from Harwich.

In the 1880s, reclaimed land that had been Ray Island was developed by the Great Eastern Railway Company (GER) as a rail depot for import/export trade with the European mainland.

The new port was named Parkeston Quay, after Charles H. Parkes, Chairman of the GER.

The existing railway line was re-routed to pass through the port – although the original railway embankment, which runs through an overgrown area known locally as The Hangings, still exists.

Parkeston is known locally as 'Spike Island' or 'Cinder City'.

From early in the 20th century, major passenger ferry services were developed, mainly to the Hook of Holland (with the slogan 'Harwich to the Hook of Holland') and later to Esbjerg in Denmark.

During both World Wars, however, Parkeston served as an important naval base.

Parkeston Quay is now named Harwich International Port and the railway station is named Harwich International.

Parkeston along with Harwich is faced, across the estuary, by the UK's busiest container port, the Port of Felixstowe in Suffolk.

The Stour Estuary

Harwich Quay – a great favourite as an observation point for visitors to obtain a close-up view of the many vessels moving in and through the estuary. These vessels range in size from canoes and yachts to the vast container ships, which can be seen using the port of Felixstowe on the opposite side of the harbour.
• 18-50mm lens at 21mm, f11, 1/500 sec, ISO 200, WB 5150

Harwich has a rich history with many sites of interest, most of which are centred on Old Harwich.

The superb natural harbour has been used since Roman times.

In Norman times the current grid layout of Harwich Town was already in existence, and it may date back further still.

The current Navyard was building ships for the Royal Navy from

1543 until 1730, when the yard was leased to private shipbuilders. It was eventually sold by the Admiralty in 1827.

Felixstowe Docks – viewed from Ha-penny Pier with Harwich Dock to the right. The Port of Felixstowe in Suffolk is the UK's busiest container port, dealing with more than 40% of the country's container cargo. It was developed following the abandonment of a project for a deep-water harbour at Maplin Sands. Recently ranked as the 28th busiest container port in the world and Europe's sixth busiest.
• 18-50mm lens at 18mm, f11, 1/400 sec, ISO 200, WB 5150

Pier Hotel – *taken from Ha-penny Pier on the old quay at Harwich*
* 55-200mm lens at 55mm, f22.0, 30 secs, ISO 100, WB 5150

Ha'penny Pier – work began on this pier in 1852 and was opened in July 1853. It was so called, funnily enough, because of the half penny toll charged (like a platform ticket).

Originally the pier was twice as long as the present one but one half burnt down in 1927.

It was a popular departure point for paddle steamers until after the First World War.

The Pier Ticket Office is a charming, typical example of late 19th century architecture. Previously having two storeys but now housing the Ha'penny Pier Visitor Centre.

The pier also accommodates the lifeboat house for the Royal National Lifeboat Institution (RNLI) inshore rescue boat, which is an Atlantic 21, B571 'British Diver II'. This boat has been on station since 31st October 1987.

Ha-penny Pier – *taken at night on the pier at the quayside in Harwich*
* 55-200mm lens at 55mm, f22.0, 30 secs, ISO 100, WB 2500

The Stour Estuary

Harwich Beach 1 *– where the beach begins
next to the dockyard*
* *18-50mm lens at 18mm, f11, 1/500 sec,
  ISO 200, WB 5150*

Harwich is best known as a ferry terminal but less well known is the town's sandy main beach. The beach is to be found near to where the *River Orwell* and *River Stour* meet and is a rare example of an emerging dune system.

Whilst Harwich Beach is not known as a sun-bathers beach it is used frequently by fishermen and sailors who launch their boats from its gently sloping sands.

Harwich Beach 2 *– with a clear view aross the
estuary to the dominant cranes of Felixtowe Docks*
* *18-50mm lens at 18mm, f11, 1/500 sec,
  ISO 200, WB 5150*

Low Lighthouse 1 – High and Low Lighthouses – there have been three pairs of lighthouses through the ages, from the original wooden pair, which no longer exist, then the brick built ones (above and below) that still stand in Harwich, to the cast iron ones (see following page) that still stand on Dovercourt seafront. None of these existing lighthouses are used for navigation purposes today.
• 18-50mm lens at 21mm, f11, 1/400 sec, ISO 200, WB 5150

Low Lighthouse 2 – these brick lighthouses were built in 1818 and were bought out by Trinity House in 1836. In 1909, the High and Low Lighthouses were sold to the Borough Council. The lower-light shown here is now used as a small museum.
• 18-50mm lens at 24mm, f11, 1/320 sec, ISO 200, WB 5150

Harwich is the headquarters of Trinity House, a unique maritime organisation, whose main objective is the safety of shipping and the welfare of seafarers.

Founded by Henry VIII In 1514, Trinity House is responsible for lights and navigational aids around Britain's coasts. It has been in Harwich since 1812 and is now the Headquarters for England and Wales, which was previously in London but moved to Harwich in 1940.

NE Essex Coast

. . . boasts one of the best climates in the country and the lowest rainfall . . .

Also known as the Essex Sunshine Coast and part of the Tendring holiday peninsula. With more than 36 miles of coastline and long sandy beaches which have received numerous Quality Coast Awards and Blue Flag Awards.

Generally, the East of England boasts one of the best climates in the country and the lowest rainfall in the UK. With the beautiful Stour Valley to the north and the *River Colne* to the south, this area is full of history, rivers, creeks, unspoilt coastlines and famous countryside.

Leading Lights 2 – *completed in 1862, they were built on rails so they could be moved to accommodate the shifting sands of Dovercourt Bay.*
• *18-50mm lens at 31mm, f22.0, 13 secs, ISO 100, WB 5500, ND Graduated Filter*

Leading Lights 1
• *18-50mm lens at 21mm, f22.0, 1/20 secs, ISO 100, WB 5150*

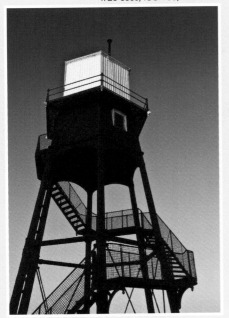

Dovercourt Bay Beach is a peaceful family resort with gentle shelving sand and shingle, Blue Flag beaches and the landmark historic leading lights. These two lighthouses, when aligned, aided navigation into the port of Harwich. They replaced the older brick built Harwich lights in 1863 because changes in the channels had made the Harwich lights misleading.

The resort enjoys panoramic views of Harwich, the port of Felixstowe and Hamford Water National Nature Reserve.

Coastal Blue – the coveted Blue Flag beach award flying above the ranks of beach huts on Dovercourt beach.
• 18-50mm lens at 21mm, f16, 1/60 sec, ISO 100, WB 5150

Coastal Gold – a coastal view looking towards The Naze at Walton from Dovercourt seafront
• 18-50mm lens at 21mm, f22.0, 1/5 secs, ISO 100, WB 5350, ND Graduated Filter

The Blue Flag Award

The award of a Blue Flag beach is based on compliance with 32 criteria covering the aspects of:

1) Environmental Education and Information

2) Water Quality

3) Environmental Management

4) Safety and Services

All Blue Flags are only awarded for one season at a time. If some of the important criteria are not fulfilled during the season the Blue Flag is withdrawn.

The NE Essex Coast is blessed with three beaches which regualarly reach the required standard: Dovercourt, Clacton and Brightlingsea.

Hamford Water and the Walton Backwaters is an area of over 2000 hectares comprising tidal creeks, mudflats, islands, salt marshes and marsh grasslands. It can be viewed best by boat or from a public footpath which runs along much of the seawall.

This area is recognised internationally and is designated as a National Nature Reserve. It is an internationally important breeding ground for Little Terns and wintering ground for Dark-bellied Brent Geese, wild foul and waders.

It also supports communities of coastal plants, which are extremely rare in Britain including Hog's Fennel, and a small population of grey and common seals.

Tide Out at Kirby Quay – the tiny port of Kirby Quay. A Thames Barge is moored opposite what used to be a granary until about 1920. This top end of Hamford Water is a popular deep water anchorage in the right conditions. In the 1800s barges loaded and unloaded a range of cargoes, including sand, gravel, chalk, lime, fertilisers, wheat and fish. The area used to have a reputation for smuggling activities.
• 10-20mm lens at 16mm, f11.0, 1/125 sec, ISO 100, WB 5150

Titchmarsh Marina – one of the most attractive harbours for yachtsmen on the East Coast.
• 18-50mm lens at 18mm, f18.0, 1/60 sec, ISO 100, WB 5150

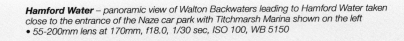

Hamford Water – panoramic view of Walton Backwaters leading to Hamford Water taken close to the entrance of the Naze car park with Titchmarsh Marina shown on the left
• 55-200mm lens at 170mm, f18.0, 1/30 sec, ISO 100, WB 5150

Beaumont Quay – *at the western end of the canal, together with a disused barn and a lime kiln, was made a scheduled ancient monument and is now under the care of the parks department of Essex County Council*
• 18-50mm lens at 33mm, f18, 1/30 sec, ISO 100, WB 5150

Beaumont Cut was a 1 kilometre (1,090 yd) long canal linking Beaumont Quay, in the parish of Beaumont-cum-Moze with Hamford Water and the North Sea. It was opened in 1832 but fell into disuse and was closed at some time in the 1930s.

The line of the canal remains watered and is easily traceable.

The coming of the railway effectively sounded the death knell for tiny ports such as these.

Arthur Ransome knew the area well. His book 'Secret Water', which has delighted generations of both children and adults, was set in the Walton Backwaters. The saltings and creeks of the Walton Backwaters make a peaceful haven from everyday life. There are narrow creeks and small islands, ideal for exploring. Stone Point on the eastern side of the Walton Channel, with its south-facing beach, is a favourite anchorage for a picnic, paddle or barbecue.

Wrecked Barge – *this wreck is the 'Rose', a 42-ton coastal barge of 1880 close to Beaumont Quay at the end of the Cut*
• 18-50mm lens at 20mm, f18, 1/25 sec, ISO 100, WB 5150

NE Essex Coast

Walton-on-the-Naze is of geological interest due to the Red Crag deposits that make up the sea cliffs, which are quite young in geological terms, being about only 1-2 million years old. They are unstable and lay on top of London Clay, about 50 million years old, which can easily be seen on the foreshore and at the base of the cliffs at low tide.

The foreshore is a good place to search for fossils from both the Red Crag and the London Clay.

The Naze Tower – *The Naze headland stretches northwards from the town of Walton on the Naze protecting Hamford Water from the North Sea. The 40m high cliffs rising to the south are unique on the Essex coastline*
  • *10-20mm lens at 10mm, f22.0, 0.6 secs, ISO 100, WB 3800*

Naze Dusk 1 – *the beach with debris from the cliffs approaching the twilight zone*
  • *10-20mm lens at 14mm, f22.0, 8 secs, ISO 100, WB 6900, ND Graduated Filter*

Naze Dawn – early morning sunrise with a once cliff top pillbox now in the sea. These WW2 reinforced concrete structures, which are all along the Essex coast, were built for troops to defend important areas such as the seashore, anti-aircraft and searchlight positions and potential landing points for airborne troops.
• 18-50mm lens at 18mm, f22.0, 0.5 secs, ISO 800, WB 5500, ND Graduated Filter

LOW LIGHT PHOTOGRAPHY

Great images can be achieved either before sunrise or after sunset by using long exposures and a tripod. Dusk is probably the time of day when the sky makes its greatest changes. Once the sun has set, the colours become warmer and a phenomenon known as 'purple light' becomes evident. This is caused by the light from the blue end of the spectrum being reflected back into the sky as in the picture below.

Also the long exposure causes a milky or blurred effect in anything moving such as the clouds adding more mood to the image.

Naze Dusk 2 – later in the the twilight zone with lights from Felixstowe docks in the background. The Naze constantly produces a range of fossil remains of animal life. The Red Crag produces shells of c. 2 million years and the London Clay holds 50 million year old sharks teeth
• 10-20mm lens at 20mm, f22.0, 30 secs, ISO 100, WB 6500, ND Graduated Filter

The old village of **Walton** is now around nine miles out to sea on the west rocks, with its old church falling into the sea in 1798.

The 'new' town grew from the early 19th century craze for sea bathing as a way to good health and, with it, came prosperity and popularity.

Today Walton still is a traditional holiday resort, and currently undergoing a new phase of exciting regeneration, aimed at maintaining its Victorian character and charm, whilst bringing it into the 21st century.

Terraced Beach Huts – viewed from the cliff above and across a small bay towards the pier these beach huts, when viewed from below, would not look out of place on a Mediterranean island
• 10-20mm lens at 20mm, f22, 1/60 sec, ISO 100, WB 5150

Beach Hut Candy – delightfully painted hut in the colours of a stick of Walton rock (candy) perhaps
• 10-20mm lens at 10mm, f22, 1/80 sec, ISO 100, WB 5150

High Tide – when the sea is really rough these plumes of seaspray can 'faithfully' resemble Icelandic geysers
• 18-50mm lens at 38mm, f22, 1/200 sec, ISO 100, WB 5150

Seaside Revival – around the beginning of the 19th Century Walton started to become a seaside resort.

As the popularity of the seaside grew and the medical benefits of sea air and salt water bathing were promoted, then the popularity of seaside towns like Walton began to rival that of the spa towns.

The coming of the railway and the building of the pier allowed travellers to make their way to Walton from the overcrowded towns and the building of hotels and other amenities gave rise to what might be described as Walton's Victorian heyday.

In recent years people have started to become interested in reviving Walton's fortunes and preserving its Victorian charm.

Unlike many developed towns Walton retains a strong flavour of its former glory days and a number of projects are already underway to improve the environment and facilities for both locals and tourists alike.

• 10-20mm lens at 10mm, f22, 1/80 sec, ISO 100, WB 5150

NE Essex Coast

Frinton-on-Sea, set between Clacton and Walton, is a non commercial seaside resort with clean sandy beaches and beach huts along the sea front. Also infamous for its greensward.

A walk in one direction brings you to Walton and The Naze and in the other direction to Holland Haven Country Park, a local nature reserve.

Close by are the villages of Kirby-le-Soken, Kirby Cross and Great Holland.

Frinton Beach – situated by the infamous greensward. When cleaned by this high tide the sandy beaches are flat and sandy - a new tapestry waiting to be freshly 'sandcastled'
- *18-50mm lens at 18mm, f22, 1/4 sec, ISO 100, WB 5150, ND Filter*

Frinton Beach Huts – on a slightly shingly beach at the end of the Greensward by the golf course on a moody September afternoon
- *18-55mm lens at 22mm, f22, 1/25 sec, ISO 100, WB 6500*

Frinton Beach Huts – viewed from the Greensward with the
Gunfleet Sands Wind Farm in the distance
• 18-50mm lens at 50mm, f22, 1/640 sec,
ISO 100, WB 5150

Frinton Greensward – the
infamous greensward which
strictly adheres to the local
bye-laws overlooking the clean
sandy beaches
• 18-50mm lens at 33mm,
f22, 1/125 sec, ISO 100,
WB 5150

Holland Haven – the boat slipway which cannot be used 1 hour either side of high tide
• 18-50mm lens at 20mm, f16, 1/60 sec, ISO 100, WB 5150

Holland Haven – situated on the coast between Clacton and Frinton with over 100 acres of unspoilt scenic coastline (ideal for bird watching). This country park is managed to conserve the landscape, coastal grazing marsh and wildlife quality of the area whilst providing for the quiet enjoyment of visitors.

In addition to some interesting breeding birds in summer, the country park is an important point for spring and autumn migrants. This Site of Special Scientific Interest (SSSI) designation recognises the rare and varied flora.

Haven Groynes – a feature of the Haven spaced approximately 70 yards apart, in need of repair or replacement
• 18-50mm lens at 18mm, f16, 1/80 sec, ISO 100, WB 5150

Pier Approach 1 – *viewed along the beach at Holland-on-Sea*
• *18-50mm lens at 20mm, f16, 1/25 sec, ISO 100, WB 5150*

Groynes – *(a feature of the Essex coastline along with the sea walls and breakwaters).*

Coastal Management – the best form of natural defence is a beach which efficiently absorbs the energy of the waves. Along many coasts, however, longshore drift causes the beach to thin out in places and erosion of the land behind becomes a problem.

Groynes are designed to slow down longshore drift and build up the beach. They are usually made of tropical hardwoods which are more resistant to marine borers and erosion. A few are made of concrete, steel or in more recent times large rocks. They are built at right angles to the shore and spaced about 50-100 metres apart. Groynes may have a life of 15-20 years but often have to be replaced rather than repaired.

Pier Approach 2 – *view of Clacton Pier with the tide coming in on the sea defences*
• *18-50mm lens at 50mm, f16, 1/15 sec, ISO 100, WB 5150*

NE Essex Coast

Clacton-on-Sea, my home town, is the largest and busiest of the Essex Sunshine Coast's resorts. It boasts beautiful golden beaches, a fun-packed pier, exceptionally maintained seafront gardens, seafront kiosks, two theatres and pleasure flights.

The Danes are back: A major feature off this part of the coast now is the Gunfleet Sands Wind Farm run by Danish firm Dong which is building three more major offshore projects. One of which will be London Array, and when completed in 2013, will be the world's biggest offshore wind farm.

West Beach 1 – viewing out to sea and the Gunfleet Sands Wind Farm
• 18-50mm lens at 50mm, f11, 1/320 sec, ISO 100, WB 5150

West Beach 2 – adjacent to Martello Bay, the Quality Coast Award West Beach has gently shelving sand with the added bonus of the operation of Clacton's pioneering child safety wristband scheme, giving parents that extra peace of mind.
• 18-50mm lens at 21mm, f16, 1/125 sec, ISO 100, WB 5150

Martello Bay located approximately one mile from the town centre and adjacent to West Beach has facilities for most water sports. Nearby is a golf course and bowling green.

The beach, which regularly boasts Blue Flag status (see page 31), is gently shelving and benefits from a child safety wristband scheme to provide reassurance for parents.

Gunfleet Sands Wind Farm – a section of the farm viewed with telephoto lens from the eastern promenade with 48 windmills generating enough electricity for 125,000 homes
• 55-200mm lens at 200mm, f11, 1/800 sec, ISO 100, WB 4500

Martello Bay – with an early morning view of the pier further along the coast. One of the two Martello Towers lies directly behind this viewpoint
• 18-55mm lens at 27mm, f22.0, 1/25 sec, ISO 100, WB 5550

Martello Tower – the second tower with the golf course and the ninth hole directly behind it. Martello towers (or simply Martellos) are small defensive forts built in several countries of the British Empire during the 19th century, from the time of the Napoleonic Wars onwards. They stand up to 40 feet (12m) high (with two floors) and typically had a garrison of one officer and 15–25 men. Their round structure and thick walls of solid masonry made them resistant to cannon fire, while their height made them an ideal platform for a single heavy artillery piece, mounted on the flat roof and able to traverse a 360° arc. A few towers had moats for extra defence. The Martello towers were used during the first half of the 19th century, but became obsolete with the introduction of powerful rifled artillery. Many have survived to the present day, often preserved as historic monuments.
• 18-55mm lens at 21mm, f22, 1/50 sec, ISO 100, WB 6000

Jaywick Sands 1 – grasses intermingling with the beach
• 18-50mm lens at 18mm, f18, 1/25 sec, ISO 100, WB 5150, ND Graduated Filter

This part of the coast from the second Martello Tower at Clacton Golf Course through to Jaywick Sands takes on a more remote and natural looking appearance with grasses and other shoreline plants intermingling with the sand and shingle.

Perhaps the making of a future dune system?

As a child, along with my young friends, who could have wished for a better playground than the NE Essex Coast.

Jaywick Sands 2 – groynes made from large rocks (see page 41)
• 18-50mm lens at 18mm, f18, 1/30 sec, ISO 100,
WB 5150, ND Graduated Filter

Jaywick Sands 3 – grasses intermingling with the beach
• 18-50mm lens at 24mm, f22, 1/50 sec,
ISO 100, WB 5150, ND Graduated Filter

St. Osyth Beach – at low tide and late evening the coastline has now become gritty and wild as it winds round to the mouth of the Colne estuary with Bradwell Power Station in the distance
• 18-50mm lens at 50mm, f18, 1/13 sec, ISO 100, WB 5150

The village of **St. Osyth** is most noted for its medieval Abbey – St. Osyth Priory.

Legend has it that Saint Osyth (or Ositha) was a young lady who was beheaded and where she fell a spring gushed from the ground. She picked up her head and walked to the door of the nunnery where she knocked three times on the door before collapsing.

To this day, Osyth's ghost walks along the priory walls carrying her head one night each year.

Mill Dam Lake – filled and emptied from St.Osyth Creek the lake is used for water skiing
• 18-50mm lens at 18mm, f18, 0.3 sec, ISO 100, WB 5150

The Boatyard – at the start of St. Osyth Creek across the road from Mill Dam Lake
just before the sun started to set
• 18-50mm lens at 23mm, f18, 0.5 sec, ISO 100, WB 5150, ND Graduated Filter

Brightlingsea Front – viewed from Mersea Island with the new development on the waterfront next to the small port and the Thames Barge 'Edme' (which is based here) anchored offshore
• 70-300mm lens at 135mm, f18, 1/80 sec, ISO 100, WB 5150

Brightlingsea is situated at the mouth of the *River Colne*, on Brightlingsea Creek.

Its traditional industries included fishing (with a renowned oyster fishery) and shipbuilding.

Essex has its very own unique 'leaning tower' in the form of Bateman's Tower.

It is adjacent to the Western Promenade with its Blue Flag status (see page 31) gently shelving sand and shingle beach.

Bateman's Tower – built in 1883 by John Bateman as a folly for his daughter to recuperate from consumption. The tower is sited on Westmarsh point at the entrance to Brightlingsea Creek on the River Colne, and is often mistaken for a Martello Tower.
• 55-200mm lens at 75mm, f16, 1/80 sec, ISO 100, WB 5150

The Port of Brightlingsea – viewed late one evening from the jetty next to the re-development of the port area and with Mersea Island in the distance
• 18-50mm lens at 26mm, f18, 5.0 sec, ISO 100, WB 6250, ND Graduated Filter

Brightlingsea is a member of the Cinque Ports, as a 'limb' of Sandwich, the only one outside of the counties of Kent and Sussex. The Cinque Ports date back over 1000 years, the original five mother ports being Hastings, Sandwich, Dover, Romney and Hythe. It is suggested that Brightlingsea oysters were a big attraction to the men of Sandwich, leading to the alliance.

The town retains an active ceremonial connection with the Cinque Ports, electing a Deputy from a guild of Freemen.

Bateman's Tower and Paddling Pool – during World War II the original roof of the folly was removed so that the tower could be used as an observation post by the Royal Observer Corps. In 2005, a restoration project by the Colne Yacht Club and funded by The Heritage Lottery Fund took place to restore the tower to its original condition, including the fitting of a replica of the original roof, refurbishing the interior of the tower and also painting the outside.
• 55-200mm lens at 88mm, f16, 1/80 sec, ISO 100, WB 5150

Thames Barge 'Edme' – featured anchored off Brightlingsea on previous page participating in the Thames Barge Race on the Blackwater Estuary. Built by Canns of Harwich in 1898, and named after the EDME Company (English Diastatic Malt Extract), situated at Mistley Quay, Essex, who are still in production today
• 55-200mm lens at 155mm, f9.0, 1/500 sec, ISO 100, WB 5150

Thames Barge 'Edme' – with little alteration since she was built and being one of only two engineless barges sailing today, she takes pride in her sailing skills. She has enjoyed considerable success in the annual Thames Barge Races, taking the annual accolade of 'Champion Barge' many times
• 55-200mm lens at 150mm, f9.0, 1/250 sec, ISO 100, WB 5150

The Colne Estuary

. . . predominantly composed of flats of fine silt with mud-flat communities . . .

The Colne Estuary is a comparatively short and branching estuary, with five tidal tributaries that flow into the main channel of the *River Colne*.

The estuary has a narrow intertidal zone predominantly composed of flats of fine silt with mud-flat communities of birds typical of south-eastern English estuaries.

Brightlingsea Seafront – viewed from East Mersea at low tide
• 18-50mm lens at 50mm, f18, 1/40 sec, ISO 100, WB 5150

Estuary View 1 – viewed from the sea wall at East Mersea at low tide
• 18-50mm lens at 28mm, f18, 1/40 sec, ISO 100, WB 5150, ND Graduated Filter

The Colne Estuary

The estuary is of international importance for a range of wintering wildfowl and waders that nest on a wide variety of coastal habitats which include mud-flat, saltmarsh, grazing marsh, sand and shingle spits, disused gravel pits and reedbeds.

Estuary View 2 – *inland view from the beach at East Mersea at low tide*
- *70-300mm lens at 180mm, f18, 1/80 sec, ISO 100, WB 5150, ND Graduated Filter*

Estuary View 3 – *inland view from the beach at East Mersea at low tide*
- *18-50mm lens at 50mm, f18, 1/50 sec, ISO 100, WB 5150, ND Graduated Filter*

The Colne Estuary

Fingringhoe Wick once a small river port serving the legionary fortress and later Roman colony at Camulodunum (Colchester).

On the evidence of Roman military equipment and coinage recovered from gravel workings on the Fingringhoe headlands, it is thought that this port was operational very early during the Roman occupation of Britain enabling direct communication with the seaports near the mouth of the Rhine.

The Old Jetty *– used for the transport of gravel from the nearby pits until 1959. The Essex Wildlife Trust acquired the pits soon after gravel extraction ended and has maintained a wide variety of habitats from the estuary foreshore to the freshwater lake and from sandy scrub to mature woodland.*
• 18-50mm lens at 18mm, f11, 1/60 sec, ISO 100, WB 5150

Wrecked Barge *– The wreck of the sailing barge 'Fly', built in Devon, pointing across the estuary to the gravel pits at Alresford*
• 18-50mm lens at 20mm, f11, 1/100 sec, ISO 100, WB 5150

Fingringhoe Wick Nature Reserve – *managed by Essex Wildlife Trust. It was once a farm, then worked as a gravel pit for forty years. In 1961 the Trust inherited this barren moonscape. The bare gravel, clay, mud and sediments are inviting seedbeds for wild plants, and today the undulating terrain is largely buried in woodland, thickets, dense scrub, ponds and a large lake. With the river frontage providing additional habitats such as saltmarsh, foreshore and inter-tidal mudflats, The Wick has become a magical place for both people and wildlife*
• 18-50mm lens at 18mm, f11, 1/100 sec, ISO 100, WB 5150, ND Graduated Filter

The Red-breasted Goose *– (Branta ruficollis) is a goose of the genus Branta. Breeds in Arctic Europe and winters in south eastern Europe. It is a rare vagrant to Britain and other western European areas, where it is sometimes found with flocks of Brent Geese although I found these amongst Greylag Geese*
• 70-300mm lens at 300mm, f5.6, 1/200 sec, ISO 400, WB 5150

COMMUNICATE WITH YOUR FELLOW MAN

Local knowledge is paramount to finding those elusive views or opportunities.

So, when out and about, talk to others who may be locals, walkers, twitchers (bird watchers) or fellow photographers.

While here at The Wick, I asked a twitcher about the best viewpoint. Afterwhich he also mentioned that some rare geese had landed at nearby Abberton Reservoir the day before. As it was on the way home I took a look, spoke to other twitchers, and, lo and behold . . .

The Colne Estuary

Alresford Creek is one of the tributaries of the estuary that reaches the sea at Brightlingsea.

The Creek is still navigable at high tide as far as Thorrington Tide Mill and provides mud berths for a number of yachts.

The nearby water-filled pits of disused workings provide pleasant lakes and homes for wildfowl and wildlife as well as the haunts of anglers.

Moored Dinghy – languishing on the mud flats at low tide
• 10-20mm lens at 12mm, f22, 1/30 sec, ISO 100, ND Graduated Filter

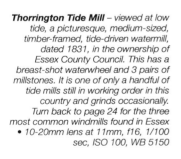

Thorrington Tide Mill – viewed at low tide, a picturesque, medium-sized, timber-framed, tide-driven watermill, dated 1831, in the ownership of Essex County Council. This has a breast-shot waterwheel and 3 pairs of millstones. It is one of only a handful of tide mills still in working order in this country and grinds occasionally. Turn back to page 24 for the three most common windmills found in Essex
• 10-20mm lens at 11mm, f16, 1/100 sec, ISO 100, WB 5150

Old Sand Jetty – *The remains of a conveyor system from the processing plant to the Creek can still be seen from the pit to the riverbank. Surrounded by current and former gravel workings, the gravel was, at one time, exported to London and beyond via sailing barges from the Creek.*
• 10-20mm lens at 14mm, f22, 1/25 sec, ISO 100, ND Graduated Filter

Landing Stage – *a number of small craft still moor alongside this ricketty protrusion*
• 10-20mm lens at 17mm, f22, 1/20 sec, ISO 100, ND Graduated Filter

SPLIT-TONING

The most successful split-tone images are those using two colours that complement each other. A good starting point is to experiment with two colours that lie opposite each other on the colour wheel. This could be red and cyan, green and magenta or, as on this page, shades of blue and yellow. Choose the darker colour for the shadows and lighter colour for the highlights.

For more subtle effects choose colours that are close to each other on the colour wheel such as red for shadows and orange or yellow for highlights.

The Colne Estuary

The history of **Wivenhoe** centres around fishing, ship building, and smuggling. The town is considered to have a bohemian quality, remaining popular with local artists, writers and (not forgetting) aspiring photographers!

Wivenhoe, which is thought to mean Wifa's Ridge, developed as a port with two prosperous shipyards and until the late 19th century was effectively a port for Colchester, as large ships were unable to navigate any further up the *River Colne*.

The town suffered significant damage when it lay close to the epicentre of one of the most destructive UK earthquakes of all time – the 1884 Colchester earthquake.

Moored Sailing Boats – *languishing on the mud flats at low tide on the other side of the Colne at Rowhedge which has also undergone redevelopment*
• *18-50mm lens at 50mm, f16, 1/50 sec, ISO 100, WB 5150*

In the 1960s, Wivenhoe Park was chosen as the location for the University of Essex. John Constable must have had a premonition for this as he painted the park in 1816.

Much of lower Wivenhoe is also a designated conservation area, with many streets being of particular architectural interest. The streets are small and quaint, leading into each other and ending at the picturesque waterfront where fishing boats and small sailing craft bob at their moorings.

Much of the quayside, has been, and continues to be very tastefully redeveloped – retaining much of the previous character of the village.

West Quay – *a section of the new development*
• *10-20mm lens at 11mm, f16, 1/60 sec, ISO 100, WB 5150*

History of the waterborne trade through the small port of Hythe on the estuary of the River Colne –
the Hythe area of Colchester, through which the Colne flows, was, from Roman times, a small port that supplied the town. The river is no longer deep enough to support ships large enough to make this viable, and water trade is now non-existent. Victorian-era plans to dredge the river ultimately failed.

Ships were able to moor in the deeper water lower down the estuary at Wivenhoe and send up smaller boats to fetch and carry merchandise. Here merchants shipped their cloth for Bordeaux or Danzig, and here they imported wine, salt, Baltic grain and continental manufactures.

Most of the boats calling here, however, were engaged in more local ventures, bringing agricultural produce or fish. The fish most in evidence at the quay was herring, either fresh or salt-cured, often shipped in by fishermen from further up the coast, but cod , plaice, rays, whiting, sprats and smelt were all commonly available. Other seafood traded here included eels and shellfish (mostly oysters, whelks and mussels).

Hythe had been built up by the then borough council as a source of income, and the publicly-owned assets here included several buildings, two cranes, and apparatus for weighing and measuring merchandise – which are now derelict.

• 10-20mm lens at 11mm, f16, 1/60 sec, ISO 100,
  WB 5150, ND Graduated Filter

Derelict Quayside Buildings – part of the quay, with its old warehouses and cranes,
which is still awaiting development
• 10-20mm lens at 12mm, f16, 1/50 sec, ISO 100, WB 5150, ND Graduated Filter

For centuries the **Hythe**, up river from Wivenhoe, was Colchester's cargo port, the destiny for many Thames barges.

The old port area is now a marina and the boats moored at the quays are homes rather than commercial vessels.

An exciting new waterside community is emerging with a multi million pound regeneration programme which will create a mixed-use development alongside the *River Colne*.

Moored Barge – on the developed part of the quay with new housing in the background
• 10-20mm lens at 10mm, f16, 1/60 sec, ISO 100, WB 5150, ND Graduated Filter

The Colne Estuary

Middlemill Weir – in the centre of Colchester, used for discharge control of the Colne, the site of a now-demolished mill
• 18-50mm lens at 18mm, f16, 1/8 sec, ISO 100, WB 5150

The tidal part of the *Colne* ends in **Colchester**, Britain's oldest recorded town, with a colourful history dating back over two thousand years. It's a history you can share by visiting the town's award-winning museums where you can see collections of international importance.

From a Roman Temple, via medieval timber frames and Victorian splendour, to cutting edge 21st century design. Panoramas of beautiful buildings reflect the town's history and contemporary appeal.

Also worth investigating are the cobbled streets of Eld Lane & Trinity Street, an area which is full of small, character-filled shops and tea/coffee rooms.

Colchester Castle – first opened to the public as a museum in 1860. Today it is still a living vibrant place. It is not only the town's flagship museum, but it is also in a real sense a symbol of Colchester, Britain's oldest recorded town.
• 10-20mm lens at 16mm, f16, 1/50 sec, ISO 100, WB 5150, ND Graduated Filter

St. Botolph's Priory 1 – the remains of one of the first Augustinian priories in England, founded about 1100. An impressive example of early Norman architecture, built of flint rubble with arches and dressings in brick – the latter mostly reused from Roman buildings nearby. Later badly damaged by cannon fire during the Civil War siege of 1648
• 10-20mm lens at 11mm, f16, 1/8 sec, ISO 100, WB 5150

St. Botolph's Priory 2 – The church was simple in design with the massive piers and arches of the nave stunning in effect. The circular piers are strengthened by triple courses of brick and the shallow pilasters running up from their capitals mark the division of the bays and the position of the roof tie-beams.
• 10-20mm lens at 10mm, f16, 1/80 sec, ISO 100, WB 5150

As well as being the inspiration for famous English artists like 18th century artist Thomas Gainsborough, 19th century landscape painter John Constable and 20th century equestrian artist Sir Alfred Munnings, the Colchester area offers cutting-edge contemporary art galleries and art cafés.

The town is also home to the University of Essex and is well known for its links with the army which has a large garrison based here.

Thames Barges at Anchor – after a days racing on the Blackwater Estuary. These are the type of boats that were able to sail up as far as Hythe Quay as featured on pages 60-61
• 55-200mm lens at 80mm, f11.0, 1/500 sec, ISO 100, WB 5150

Chappel Viaduct

... originally planned that it should cross the Colne Valley on a timber viaduct ...

OK folks! A quick trip inland to view the viaduct crossing the *River Colne* further upstream and the Colne Valley. An awe-inspiring landmark of such immensity that it never fails to impress.

It is the largest viaduct in East Anglia, and was opened in 1849, after less than three years construction, despite containing seven million bricks.

With thirty two arches of thirty foot span and a total length of 1066 feet, it stands seventy five feet at its maximum height above the valley it crosses.

It was originally planned that it should cross the Colne Valley on a timber viaduct about 70ft high.

In the event, good brick earth was subsequently discovered at Mount Bures, so it was decided to change to brick arches which would be cheaper to maintain. In July 1847 after two million bricks had been readied and a workforce of approximately 600 men assembled, work began on the foundations.

Chappel Viaduct 1 – *viewed from open farmland*
• 55-200mm lens at 160mm, f22.0, 1/25 sec, ISO 100, WB 4500

Chappel Viaduct 2 – view from the A1124 early one misty October morning with the sleepy village of Wakes Colne seen through the arches
• 150-500mm lens at 150mm, f22.0, 1/30 sec, ISO 100, WB4850

Most of the men working on the viaduct would have been encamped locally, possibly around what is now known as the Chappel Millennium Green. The bulk of them were East Anglian farm workers, desperately in need of some form of better paid employment. They worked with phenominal efficiency and the foundations having been completed in February 1848, the viaduct was finished except for the parapets by the following February.

Some 5 or 6 million bricks are presumed to have been used to complete with the piers constructed hollow to save money.

Chappel Viaduct 3 – view of the hollow piers
• 10-20mm lens at 10mm, f22.0, 1/4 sec, ISO 100, WB 4150

Chappel Viaduct

Chappel Viaduct 4 *– early morning misty view in October as the sun rises above the parapet*
* *55-200mm lens at 90mm, f22.0, 1/125 sec, ISO 100*

Chappel Viaduct 5 *– view from beneath the arches at around midday*
* *55-200mm lens at 165mm, f22.0, 1/8 sec, ISO 100, WB 4700*

A further significant feature of the viaduct is that it is on a gradient and the Sudbury end is 9 feet 6 inches higher than the Marks Tey end.

The total cost estimated by the engineer Peter Bruff was £21,000, which seems

Chappel Viaduct 6
– wide angled view from beneath the arches
* *10-20mm lens at 10mm, f22.0, 1/10 sec, ISO 100, WB 5000*

Chappel Viaduct 7 – view from open farmland after the harvest
• 55-200mm lens at 70mm, f22.0, 1/25 sec, ISO 100, WB 5000

very inexpensive when compared to the present day cost of, for example, having a small extension built to one's home.

The Viaduct is now a listed monument and, despite once being threatened by the Beeching railway cuts, still carries trains across the *River Colne* and valley.

DYNAMIC DIAGONALS

Diagonal lines are a great way of leading the viewer into the picture. When they lead from the edge of the frame towards the centre or a significant part of the scene they can help to concentrate the attention onto this area of the image.

With the above picture the point of focal attention is a point on the 'rule of thirds' where the converging furrows on the field meet the viaduct. The diagonals lead you into it and then out top right of the frame.

In this case by breaking the rules and deliberately twisting the frame you can produce dynamic diagonals to boost the picture.

Chappel Viaduct 8 – the single carriage train at present runs approximately every hour
• 150-500mm lens at 267mm, f22.0, 1/125 sec, ISO 400, WB 5000

Mersea Island

... one of the most important geological sites in Essex ...

Mersea Island is located in the estuary of the rivers *Colne* to the north and *Blackwater* to the south. Approximately nine miles south-east of Colchester and is the most easterly inhabited island in the UK.

It is linked to the mainland by The Strood, an artificial causeway about half a mile long, which is liable to flooding at high tide.

Its area of five square miles includes the village of East Mersea and the town of West Mersea.

London Clay – the exposed London Clay of the desolate eroding terrain
• 55-200mm lens at 55mm, f22.0, 1/4 sec, ISO 100, WB 5150

Sticky Defences – situated at East Mersea a polder scheme has been installed on the mud flats to encourage a build up of silt to help salt marsh vegetation to colonise and hence prevent high tides eroding the base of the nearby cliffs at Cudmore Grove Country Park
• 18-50mm lens at 50mm, f18, 0.6 sec, ISO 100, WB 5150

East Mersea Beach – viewed at the approach to Cudmore Grove Country Park
• 18-50mm lens at 18mm, f18, 1/50 sec, ISO 100, WB 5150, ND Graduated Filter

Ancient Beach – the eroding terrain at
Cudmore Grove Country Park with disintegrating
pillbox in the foreground that was once on the cliff
• 18-50mm lens at 21mm, f18.0, 1/10 sec,
ISO 100, WB 5150, ND Graduated Filter

East Mersea is one of the most important geological sites in Essex, and the cliffs at Cudmore Grove provide superb exposures of Thames/Medway gravels laid down during the Hoxnian interglacial stage when monkeys, bears and early man lived in Essex.

Beneath the beach gravel, and inaccessible, are channel deposits from the more recent Ipswichian interglacial stage which have yielded bones of hippopotamus, elephant, rhinoceros and hyena.

Mersea Island

Mersea has always been an important player in the oyster market, bringing in the famed 'Mersea Native Oysters'. Even now, oysters are still one of the main industries of the area, along with fishing, farming, boat-building, yacht design and sailmaking.

West Mersea is very much a tourist area. Along with the beach huts you can wander along the Hard where boats are stored in large boat parks.

Mersey Beach Huts – located on West Mersea beach one spring afternoon
• *10-20mm lens at 11mm, f22.0, 1/8 sec, ISO 100, WB 5350, ND Graduated Filter*

Moody Beach Huts – located on West Mersea beach one overcast afternoon
• *10-20mm lens at 12mm, f22.0, 1/15 sec, ISO 100, WB 5150, ND Graduated Filter*

Mersea Island

There is a thriving sailing community with a yearly regatta which comes at the end of a week of hard racing.

During Mersea Week (as it is known) there is a round-the-island race, though you cannot actually sail the entire way round the island due to The Strood. So waiting are hundreds of trailers and helpers ready to carry boats across the road!

Hanging Buoys – *adorning the underside of the yaught club*
• *18-50mm lens at 50mm, f8.0, 1/40 sec, ISO 100, WB 5150*

Derelict Boat – *plenty of these to photograph at West Mersea*
• *18-50mm lens at 28mm, f8.0, 1/50 sec, ISO 100, WB 5150*

Mersea Island

Also along the Hard and between the coast road at Besom Fleet are the sand dunes with many upturned dinghies interspersed with long grasses.

Also here is a large house boat community adjoining the boat parks.

The beach surrounding the island is mainly sand and shingle, turning into mudflats as you wade further out.

Sand Dunes – interspersed with sailing boats and upturned dinghies
• 10-20mm lens at 11mm, f16.0, 1/25 sec, ISO 100, WB 5150

Mersey House Boats 1 – plenty of these to photograph at West Mersea
• 10-20mm lens at 13mm, f16.0, 1/30 sec, ISO 100, WB 5150, ND Graduated Filter

Mersey House Boats 2 – plenty of these to photograph at West Mersea with the surrounding community of houseboat residents and the close proximity of facilities and leisure. Such as the pubs and restaurants, including the famous Company Shed renowned for its oyster cuisine, making life here better than could be imagined.
• 10-20mm lens at 10mm, f16.0, 1/20 sec, ISO 100, WB 5150, ND Graduated Filter

Mersea Island

Thames Barge Racing – *off the Mersea coast*
- *55-200mm lens at 55mm, f9.0, 1/400 sec, ISO 100, WB 5150*

Why are all the sails that red colour?

Sail dressing of yester year often had 'secret' ingredients. To make these Thames barge sails more efficient and to prolong their life they were dressed with a mixture of oil (traditionally fish oil), seawater (and/or horse urine if available!) and red ochre.

The red ochre was used as a colouring agent, without which the sails would look a dirty grey colour.

Tollesbury Waterside

. . . has for centuries, relied on the harvests from both the land and the sea . . .

The village of **Tollesbury** sits on a small peninsula, to the north is Tollesbury Fleet and Old Hall marshes, to the south the *River Blackwater*. The waterside makes the area popular with bird spotters, walkers and sailors.

Because of it's geographical situation Tollesbury has for centuries, relied on the harvests from both the land and the sea and the village has become known as 'The Village of the Plough and Sail'.

The Lightship 'Trinity' 1
• *10-20mm lens at 11mm, f16, 1/15 sec, ISO 100, WB 5150, ND Graduated Filter*

The Lightship 'Trinity' 2 – Fellowship Afloat's centre, a charity organisation providing outdoor pursuits for young people. Converted from a lightvessel in 1990, she can accommodate up to 36 guests.
Trinity is moored in Woodrolfe Creek, a sheltered inlet on the north side of the River Blackwater estuary and the main channel through the marshes into the Marina at low tide.
The sea wall in the background overlooks the salt marshes towards Tollesbury Fleet and Old Hall Marsh. Salt marshes such as these support a diverse range of plants which can tolerate high salt concentrations. These can produce large areas of colour in the summer such as areas of pink Thrift and drifts of purple Sea Lavender.
• *10-20mm lens at 15mm, f16, 1/20 sec, ISO 100, WB 5150, ND Graduated Filter*

Moorings in Woodrolfe Creek – *at low tide with Tollesbury marina in the background*
* *10-20mm lens at 20mm, f16, 1/15 sec, ISO 100, WB 5150, ND Graduated Filter*

Tollesbury Marina – *part of, at low tide*
* *18-50mm lens at 50mm, f16, 1/13 sec, ISO 100, WB 5150, ND Graduated Filter*

The marina at Tollesbury is a tidal harbour with access for about 2 hours either side of high water depending on the state of the tide, weather conditions and the draft of your boat.

It is open to visiting boats for a daily fee.

Tollesbury Waterside

Tollesbury Sail Lofts were built at the end of the nineteenth century (c1890) to serve the local fishing fleet and probably the great J class yachts, owned by wealthy Edwardians, which were skippered around the Mediterranean by the local men from or around Tollesbury.

The Boatyard – *rowing boats left high and dry at low tide*
• *18-50mm lens at 50mm, f16, 1/50 sec, ISO 100, WB 5150, ND Graduated Filter*

Tollesbury Sail Lofts – *built during the late 1800's, these buildings were used to store the sails from the local fishing fleet and, during the winter months, the great J class yachts which were owned by wealthy Edwardians and skippered around the Mediterranean by men from Tollesbury. The most well known of these yachts, 'Flying' was entered by Tommy Sopwith in the Americas Cup.*
Despite these buildings being a quarter mile from the waterside these lofts are built off the ground. Being on the seaward side of the sea wall they are affected by high spring tides.
• *18-55mm lens at 28mm, f22, 1/160 sec, ISO 400, WB 5250*

Heybridge Basin

... used as a haven for visiting yachts or as a starting point to explore the waterway ...

The Front Sea Lock – *where canal meets estuary*
- *18-50mm lens at 50mm, f22.0, 1/40 sec, ISO 100, WB 6700, ND Graduated Filter*

The village of **Heybridge Basin** owes its existence to *The Chelmer and Blackwater Navigation* built between 1793 and 1797 which links Chelmsford with the *Blackwater* estuary.

Now used as a haven for visiting yachts or as a starting point to explore the waterway either afloat or on foot along the towpath which is a public right of way to Chelmsford.

Heybridge Lock was built as the sea terminus for The *Chelmer and Blackwater Navigation*. Heybridge Basin grew up around the lock as there was plenty of work unloading the ships or Thames barges into the canal barges. The village has continued to expand into the 21st century despite the cessation of commercial activities on the canal. This expansion has continued because of the excellent sailing facilities on the *River Blackwater* and also because

of the 150-200 yachts permanently moored in the Basin. The village also has an excellent tea room run by the world famous nearby Tiptree Jam Company who have recently celebrated 125 years.

There is (during the summer) an hourly ferry service from the Basin to the shops in Maldon run by Basin Pleasure Boats or you can take your own dinghy up to the shops. Small craft can navigate the 14 miles upstream to Chelmsford (by arrangement).

The Wharf – constructed on a mound known as Lock Hill with the
Thames Barge 'Decima' moored behind the tea rooms
• 18-50mm lens at 21mm, f22.0, 1/5 sec,
ISO 100, WB 6500, ND Graduated Filter

NEUTRAL DENSITY (ND) GRADUATED FILTERS

It is said that there is no longer any need to use filters as these effects can easily be produced in digital image-editing software at post production stage.

However, the ND Graduated filter is one of a few filters that are still best used on camera. When photographing landscapes there can often be a three or four stop difference between the light reading for the foreground and the sky. If you expose for the sky, the foreground will appear dark, but if you expose for the foreground, the sky will appear burnt out.

This graduated filter corrects this imbalance. The filter is half grey/half clear and by ensuring the darker part filters out the lighter part of the scene the exposure balance is restored.

Basin Sunrise – casting a glow over the mudflats of
the River Blackwater estuary at low tide
• 18-50mm lens at 23mm, f22.0, 1/5 sec, ISO 100,
WB 6300, ND Graduated Filter

Heybridge Basin

Inner Lock Gate – early morning wide angle view of the marina
• 10-20mm lens at 10mm, f22.0, 1/50 sec, ISO 100, WB 6500

The sea lock and moorings before Hall Bridge on the Navigation are in many respects comparable to an estuarial marina.

Hall Bridge (see next chapter) is the first on the waterway, heading upstream, and forming the dividing line between the 'marina' and the inland waterway itself.

Last Mast – last moored yacht before reaching Hall Bridge
• 18-50mm lens at 26mm, f22.0, 1/15 sec, ISO 100, WB 5500

Mooring Platform – a wide angle view of the estuarial marina/canal at Heybridge
• 10-20mm lens at 10mm, f22.0, 1/13 sec, ISO 100, WB 6700

Yellow Iris – handsome display on the canal near Hall Bridge. Also known as the Yellow Flag Iris, it is a true aquatic plant that is native to England. It is the only yellow-flowered iris and reaches up to 48" in the wild.
• 55-200mm lens at 55mm, f8.0, 1/80 sec, ISO 100, WB 6700

Heybridge Basin

• 55-200mm lens at 55mm, f9.0,
1/125 sec, ISO 100, WB 5150

Tiptree Jams and Preserves not only run a tea room at Heybridge but are also, currently, sponsors of the Thames Sailing Barge 'Decima' which is based here.

She is a regular sight (as are the other barges) throughout the Essex coast and regularly competes in the series of barge match races held annually - with many first places to her credit! Now for hire on the *River Blackwater* or throughout the East Coast.

Built in 1899 (one of 20 identical steel barges) by F. G. Fay & Co. of Southampton.

'Decima' is in her 113th year and, after numerous mishaps during her lifetime, is still sailing the same waters that she always has since 1899!

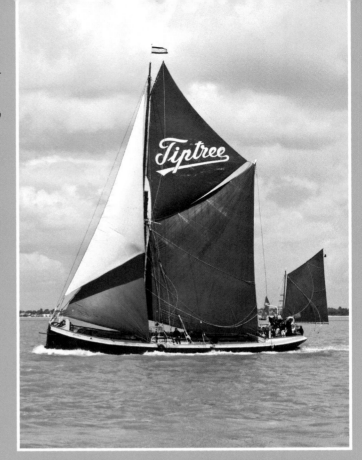

• 55-200mm lens at 55mm, f9.0,
1/400 sec, ISO 100, WB 5150

Almost all Thames barges were built purely as sailing craft, without any other form of power.

Most of the barges afloat today do not rely solely on the wind and have fairly large auxiliary diesel engines of 5-7 litres or so.

There are at least two still sailing regularly today which have never been fitted with an engine.

• 55-200mm lens at 140mm, f9.0, 1/400 sec, ISO 100, WB 5150

The Chelmer and Blackwater Navigation

. . . natuaral paradise . . . consisting of locks, weirs and cricket bat willows . .

The Chelmer and Blackwater Navigation is today a natural paradise for walkers, boaters and anglers consisting of locks, weirs and cricket bat willows. Once busy with horse-drawn barges the 14 miles of towpath offer recreational walking through beautiful Essex scenery.

Constructed in 1793 when a canal was dug from the estuary for one and a half miles across the marshes to the village of Heybridge. Here the canal joined the *River Blackwater* and was linked, via a short artificial cut at Beeleigh, with the *River*

Natural Paradise – delightful scene of lily pads and blooming hawthorn on the Long Pond stretch of the canal at Maldon • 18-50mm lens at 31mm, f22.0, 1/20 sec, ISO 100, WB 6200

View from Hall Bridge – the first bridge on the waterway, heading upstream, and forming the dividing line between the 'marina' at Heybridge and the inland waterway itself. The image converges on Wave Bridge in the distance along the Long Pond • 10-20mm lens at 20mm, f22, 1/8 sec, ISO 100, WB 6000

The Chelmer and Blackwater Navigation

Canal at Maldon – *view of the canal in front of an office building as it passes through Maldon*
- *10-20mm lens at 18mm, f22, 1/10 sec, ISO 100, WB 5500*

Chelmer and then on to the town of Chelmsford ten miles inland.

This section of the canal, from Heybridge Basin to Beeleigh, is known as the Long Pond. For most of its length the towpath is designated a bridleway which then becomes a footpath.

There are twelve locks in all (roughly one mile apart) and the following pages take you along the navigation from Heybridge Basin (see previous pages) through to Springfield Basin in Chelmsford.

Bridge over Tranquil Water 1 – *under the A414 road bridge at Maldon. A composition of modern and bygone infrastructure in a monotone and full colour combination*
- *10-20mm lens at 20mm, f22, 1/15 sec, ISO 100, WB 5500*

The Chelmer and Blackwater Navigation

Approach to Chapman's Bridge – *on the Long Pond, here the bridalway ends and becomes a footpath although considerate cyclists are not discouraged*
• *10-20mm lens at 11mm, f22.0, 1/80 sec, ISO 400*

Chapmans Bridge on the outskirts of Maldon is a grade II building listed as a:
Canal accommodation bridge. Early 19th century. Cream Gault brick with red-brick curving string band beneath each parapet. Elliptical or deformed semicircular arch and slightly projecting piers at each approach.

Accommodation bridges were built over canals to preserve ancient rights of way and typically provided a convenience for farmers, who required access to their fields that lay on the opposite side of the canal.

Chapman's Bridge – *one of a number of the original over 200 year old brick (humpback) bridges in good repair along the canal. Near to Elms Farm Park, created in the mid 1990's as a new country park as part of a large housing development on the edge of Heybridge. The site is quite diverse consisting as it does of meadows, thick vegetation, scrub, trees, wet ditches and open water. Their is a man-made lake and island which makes up about one third of the site.*
• *18-50mm lens at 20mm, f22, 1/100 sec, ISO 400, WB 6000*

Flood Gates – a view of open flood gates looking back in the direction
of Chapman's Bridge and Heybridge
• 10-20mm lens at 13mm, f22.0, 1/25 sec, ISO 100, WB 6000

The canal path now
runs into a tranquil
phase before
passing by the
boundary of Maldon
Golf Club on its way
to the flood gates of
Beeleigh Lock.

Footbridge and Flood Gates
– John Rennie's picturesque
redbrick bridge and open flood
gates at Beeleigh
• 10-20mm lens at 10mm,
f22.0, 1/15 sec,
ISO 100, WB 5500

The Chelmer and Blackwater Navigation

Langford Canal – early morning view where it meets the River Blackwater with Long Weir and footbridge in the distance. Langford Cut was dug in 1793 (4 years prior to the C&B). It was used to transport wheat from a mill in Langford to the river. The last barge to use the canal was in 1881
• 10-20mm lens at 18mm, f22.0, 0.6 sec, ISO 100, WB 5800, ND Graduated Filter

Long Weir – early morning view of the River Blackwater where it brushes with the River Chelmer overflowing into it
• 10-20mm lens at 12mm, f22.0, 1/13 sec, ISO 100, WB 6000, ND Graduated Filter

The Chelmer and Blackwater Navigation

Here at Beeleigh, the now derelict Langford Cut joins the Long Pond, which is fed from the junction of the Chelmer and Blackwater rivers, whose surplus water flows into a channel over the Long Weir.

On very high tides, the flow can reverse, so it is important to ensure that the flood gates below John Rennie's picturesque redbrick bridge (see previous page 91) are left closed.

Beeleigh Lock 1 – evening view looking back to John Rennie's footbridge and the Long Weir where the Chelmer overflows into the Blackwater
• 10-20mm lens at 10mm, f22.0, 1/15 sec, ISO 100, WB 6200

Beeleigh Lock 2 – a view looking towards Rickett's Lock and Chelmsford taken at sunset
• 18-50mm lens at 20mm, f22.0, 1.3 sec, ISO 100, WB 6700, ND Graduated Filter

The Chelmer and Blackwater Navigation

Hawthorn in Bloom – looking back from the bridge towards Beeleigh
• 10-20mm lens at 20mm, f22.0, 1/4 sec, ISO 100, WB 6200

The canal continues with the winding, wooded compound between Beeleigh and Rickett's locks. The land between Rickett's Lock and its weir is abundant with wild flowers and insects, in particular, dragonflies and damselflies. Here also, as with the rest of the navigation, is excellent and well stocked coarse fishing.

Approaching Rickett's Lock – view of the bridge as we approach the lock
• 55-200mm lens at 75mm, f22.0, 1/30 sec, ISO 100, WB 6200

Rickett's Lock – a wide angle view of Rickett's Lock and Bridge and (above) views approaching the bridge and looking back from the bridge
• 10-20mm lens at 10mm, f22.0, 1/30 sec, ISO 100, WB 6000

Banded Demoiselle Damselfly – Banded Demoiselle
(Calopteryx splendens) Damselfly (Zygoptera)
• 150-500mm lens at 500mm, f9.0, 1/320 sec,
ISO 400, WB 4850

The Chelmer and Blackwater Navigation

Banded Demoiselle Damselfly – *male Banded Demoiselle*
(Calopteryx splendens) Damselfly (Zygoptera)
- *150-500mm lens at 500mm, f9.0, 1/125 sec, ISO 400, WB 4850*

Essex Skipper –
(Thymelicus lineola)
- *150-500mm lens at 500mm,*
  f9.0, 1/640 sec, ISO 400,
  WB 4450

Banded Demoiselle Damselfly – *female Banded Demoiselle*
(Calopteryx splendens) Damselfly (Zygoptera)
- *150-500mm lens at 500mm, f9.0, 1/250 sec, ISO 400, WB 4850*

The navigation on its way to Hoe Mill Lock, with its cricket bat willows and watermeadows, becomes very like 'Constable Country' in Dedham Vale, which indeed it is, as the famous artist's family milled not only on the Stour, but also at Hoe Mill, near the lock.

Also, near to where a water pipe crosses the Navigation just before reaching Hoe Mill Lock is Sugar Baker's Holes, a pioneering site for the extraction of sugar from beet, established in 1832. Here, also, are pretty canalside cottages that were once used by the factory workers.

Scarce Chaser Dragonfly –
female (Libellula fulva)
• 150-500mm lens at 500mm,
f6.3, 1/800 sec,
ISO 400, WB 5150

Scarce Chaser Dragonfly –
male (Libellula fulva)
• 55-200mm lens at 200mm,
f11.0, 1/640 sec,
ISO 400, WB 4350

The Chelmer and Blackwater Navigation

Hoe Mill Lock *– with moorings against the non-towpath bank*
• 18-50mm lens at 40mm, f22.0, 0.3 sec, ISO 100, WB 6200

Hoe Mill Lock, the deepest on the waterway, is home to the first of the Navigation's four main inland mooring areas, with about forty boats lying against the non-towpath bank. We are also near to the only remaining original inland lock house.

Just above the moorings the towpath crosses an interesting horse bridge.

A short distance above Hoe Mill, also on the far bank, lies tiny All Saints Church, Ulting, perhaps the most photographed location on the river.

Hoe Mill Boats *– moorings against the non-towpath bank*
• 18-50mm lens at 31mm, f22.0, 1/5 sec, ISO 100, WB 6200

All Saints Church – near Hoe Mill Lock at Ulting. There has been a church here since about 1150AD. The siting of the church suggests that the present site may have been an early Christian meeting place in Saxon times due to interesting 'parch marks' appearing in the churchyard during the dry spring of 2007 which could be the lines of earlier structures
• 18-50mm lens at 24mm, f22.0, 1/8 sec, ISO 100, WB 6000, ND Graduated Filter

Rushe's Lock Approach – *from the direction of Hoe Mill Lock*
• *55-200mm lens at 60mm, f22.0, 1/15 sec, ISO 100, WB 5500*

A few bends beyond brings remote Rushe's Lock (named after a local farming family) and its lush surroundings into view.

Idyllic in its pastoral setting under the rise of Danbury Hill, this is, in my opinion, one of the two most picturesque of the twelve locks with its adjacent weir.

Definition of a Weir: *a small overflow dam used to alter the flow characteristics of a river or stream. In most cases weirs are a form of barrier crossing the river causing water to pool behind the structure but allowing water to flow over the top. Weirs are commonly used to alter the flow regime of the river, prevent flooding, measure discharge and to help render a river navigable.*

Weir at Rushe's Lock – picturesque
• 10-20mm lens at 20mm, f22.0, 1/25 sec, ISO 100, WB 6000

Lock Jaws – close up of the approach to Rushe's Lock
• 10-20mm lens at 11mm, f22.0, 1/8 sec, ISO 100, WB 6000

The Chelmer and Blackwater Navigation

Cricket Bat Willows – from now on we can only admire the four willow-lined miles all the way to sleepy Sandford.

Essex has long been renowned for its willow trees. Famous around the world for the manufacture of some of the best 'test match standard' cricket bats.

Due to being a low lying county with numerous rivers, wide flood plains and a temperate climate, the cricket bat willow, (Silex alba coerulea), a hybrid willow species, thrives in Essex but is much more difficult to grow in other parts of the country.

The tradition of growing and harvesting cricket bat willows, grown specifically for the production of cricket bats, along the navigation is one which has helped the landscape develop some of it's unique characteristics, both visually and ecologically.

The industry ensures the continuation of this species of tree in the landscape by actively farming and managing the trees as a crop to achieve the optimum potential when harvested. Trees of 56 inch circumference (around 15 years old) are felled and new trees planted alongside old nearly every year.

Willows have been a significant part of our Essex landscape for over 100 years, so, lets do our bit by encouraging our schools and children to play more cricket!

Evening Sky – *looking back towards Rushe's Lock from the footbridge between Rushe's and Paper Mill Locks with the regimental lines of Cricket Bat Willows*
• 18-50mm lens at 18mm, f22.0, 0.5 sec, ISO 100, WB 6500

Willows Bend 1 – *looking back towards the footbridge between Rushe's and Paper Mill Locks*
• 18-50mm lens at 18mm, f22.0, 1/10 sec, ISO 100, WB 6700

Willows Bend 2 *– looking forward towards Paper Mill Lock later in the evening at sunset*
• 55-200mm lens at 60mm, f4.5, 1/13 sec, ISO 100, WB 6200, ND Graduated Filter

Paper Mill Bridge – moored boats on the canal by Paper Mill Lock looking towards the bridge. Paper Mill was the first place in Essex where paper was produced. In 1792 there were two mills here, one for grinding corn, the other for pulp rags. Paper Mill Lock is now the headquarters of the Chelmer and Blackwater Navigation company.

• 18-50mm lens at 40mm, f22.0, 1/6 sec, ISO 100, WB 6000

Soon after passing a picturesque waterside cottage called Smuggler's Barn, Paper Mill is glimpsed - framed by a road bridge.

This is the busiest of the mooring areas, with over one hundred boats, ranging from canoes and dinghies, through to modern and classic cruisers, to an increasing fleet of traditional narrowboats.

Two of the waterway's four passenger boats operate from Paper Mill. There is also the popular lock side tearoom where you can also hire out dinghies and canoes.

The tearoom occupies the former stables where horses spent the

Weir at Paper Mill Lock – moored boats by the weir
• 55-200mm lens at 75mm, f22, 1/30 sec, ISO 100, WB 5050

night, whilst opposite is a little red brick shelter where the bargees slept. This is now used as the Navigation's office.

Paper Mill Lock – looking towards Baddow Mill Lock one fine spring evening
• 18-50mm lens at 50mm, f22.0, 1/6 sec, ISO 100, WB 6700

The Chelmer and Blackwater Navigation

Willow Reflection *– looking back towards Paper Mill Lock one fine autumn morning*
* *10-20mm lens at 11mm, f22.0, 1/6 sec, ISO 100, WB 6000*

The willow lined waterway pleasantly runs on through to Little Baddow.

Along this stretch as with many stretches of the navigation you can (if alone) aimlessly ponder, reminisce, or take account of ones self.

On a nice day the feeling of remoteness, peace and tranquility is second to none.

Willow Coppicing *– A stretch of the canal near Little Baddow where the cricket bat willow or White Willow (Salix alba) is being managed*
* *18-50mm lens at 20mm, f22.0, 1/25 sec, ISO 400, WB 5500, ND Graduated Filter*

Baddow Mill House – next to the weir
• 55-200mm lens at 67mm, f22.0,
 1/10 sec, ISO 100, WB 6700

After passing over a road bridge you approach a footbridge via a magnificent avenue of towering willows.

Afterwhich, the other favourite picturesque lock of mine at Baddow Mill awaits. (See also introduction picture, p5)

Baddow Mill Lock 2 – delightful view of the lock taken one spring evening
• 18-50mm lens at 23mm, f22.0, 1/40 sec, ISO 100, WB 6700

The Chelmer and Blackwater Navigation

Dog Rose – *(Rosa canina), wild rose, symbol of the Tudor monarchs, fruit is the red rose hip, from which the syrup is made*
• *150-500mm lens at 500mm, f9.0, 1/200 sec, ISO 400, WB 4700*

As well as the cricket bat willows the navigation hosts a wealth of natural vegetation.

Spring is when the canal is at its most splendourous with the hawthorn and canal side plants in full bloom.

For centuries the fast growing, sturdy hawthorn has divided fields, also known as May or Quickthorn.

Cow Parsley Floret – *flowerhead of canal side Cow Parsley (Anthriscus sylvestris)*
• *18-50mm lens at 18mm, f2.8, 1/500 sec, ISO 400, WB 7500*

Musk Thistle – *(Carduus nutans) with its musky odour*
• *150-500mm lens at 500mm, f7.1, 1/200 sec, ISO 200, WB 4900*

Hawthorn Display – (Crataegus monogyna) sturdy hawthorn has divided fields across mile upon mile of British countryside
• 18-50mm lens at 50mm, f2.8, 1/500 sec, ISO 400, WB 6700

Hemlock – (Conium maculatum) one of the tallest members of the parsley family. Socrates was poisioned by hemlock and the plant formed part of the witches brew in Shakepeare's 'Macbeth'
• 150-500mm lens at 500mm, f9.0, 1/800 sec, ISO 400, WB 4800

Lonesome Willow – *midway between Baddow Mill
and Stoneham's Locks*
• *55-200mm lens at 55mm, f22.0, 1/10 sec,
ISO 400, WB 5150, ND Graduated Filter*

Stoneham's Lock – a view of the recently repaired lock with duotone effect
• 10-20mm lens at 11mm, f22.0, 1/5 sec, ISO 100, WB 5500, ND Graduated Filter

The navigation, picturesque and isolated, has recently been rescued from bankruptcy and is now run by The Inland Waterways Association.

IWA has already carried out a great deal of work (like here at Stoneham's lock and weir) and hopes to improve the state of the navigation further in forthcoming years.

A local treasure well worth maintaining for the present and future generations to enjoy.

Weir at Stoneham's Lock 1 – viewed from behind the weir next to the lock
• 18-50mm lens at 38mm, f22.0, 0.6 sec, ISO 100, WB 6200

Weir at Stoneham's Lock 2 – a view of the weir next to the lock
• 18-50mm lens at 50mm, f22.0, 1/5 sec, ISO 100, WB 6700, ND Filter

Noisy Corner – *the point where the canal bends and runs parallel with the busy A12 on the right*
• *18-50mm lens at 26mm, f22.0, 1/40 sec, ISO 100, WB 6200*

Weir at Cutons Lock 2 – *close up view of the weir in Spring*
• *18-50mm lens at 50mm, f22.0, 1/10 sec, ISO 100, WB 6200, ND Filter*

Weir at Cutons Lock 1 – *approach view to the left of the lock*
• *55-200mm lens at 60mm, f22.0, 1/15 sec, ISO 100, WB 6700*

Cutons Lock *– view of the lock one sunny spring morning*
• *18-50mm lens at 18mm, f22.0, 1/8 sec, ISO 100,*
WB 5500, ND Graduated Filter

Just before Cuton Lock and weir and beyond is the noisiest part of the canal because of its close proximity to the busy A12 carriageway.

However, on the whole, the navigation runs through a largely unspoilt part of rural Essex.

Designed by the renowned canal engineer John Rennie and completed on 3rd June 1797, most of his original structures are intact, together with the entire length of the original waterway.

Weir at Cutons Lock 3 *– close up view of the weir in September*
• *18-50mm lens at 50mm, f22.0, 1.3 sec, ISO 100, WB 5500, ND Filter*

The Chelmer and Blackwater Navigation

The willow lined waterway continues on, leaving the noisy A12 behind, and here at Sandford there are over twenty moorings in the lock cut between two beautiful hump-backed bridges.

The small passenger boat *Blackwater Rose* is based here.

One of two remaining lighters, the historic timber-built *Susan* is based at the museum at Sandford Mill, while the steel-built *Julie* still works on the waterway as a maintenance craft.

The Canal at Sandford Mill 2 – with moored boats between the lock gates and
Bundock's Bridge in the distance
• 10-20mm lens at 11mm, f22.0, 1/5 sec, ISO 100,
WB 6050, ND Graduated Filter

Bundock's Bridge – a view to
Sandford Mill Lock in the
distance from under the bridge
• 18-50mm lens at 28mm,
f22.0, 1/6 sec, ISO 100, WB
6000, Camera Flash

The Chelmer and Blackwater Navigation

After Bundock's Bridge the navigation takes a right hand turn with lush vegetation on the left.

As we approach the next lock at Barnes Mill the towpath on the right, as in many places along the canal, is separated from the bank by swathes of riverside plants including Cow Parsley.

Barnes Mill, now converted to residential use, overlooks the mill pool below the lock . . .

Weeping Willow Trees *– on the bend in the canal just after Bundock's Bridge*
• 18-50mm lens at 50mm, f22.0, 1/5 sec, ISO 100, WB 6000

Barnes Mill *– the dwellings and Mill Pond*
• 18-50mm lens at 43mm, f22.0, 1/8 sec, ISO 100, WB 6200

Footbridge at Barnes Mill 1 – a view of Barnes Mill and the Mill Pond can be seen from this bridge to the left
• 18-50mm lens at 20mm, f22.0, 1/15 sec, ISO 100, WB 6700

CREATIVE FILTER EFFECTS

Most landscape photographers have an array of necessary filters to cope with the conditions such as polariser, neutral density and graduated filters which are fitted to the front of the lens.

But, there are also a host of creative filters on the market such as a sepia tone and a centre spot diffuser as used for the image below to create a mystical effect.

Footbridge at Barnes Mill 2 – a view of the footbridge that spans the entrance to Barnes Mill and the Mill Pond from the other side of the canal in heavy sepia tone
• 55-200mm lens at 67mm, f5.6, 1/60 sec, ISO 100, creative filters x 2

The Chelmer and Blackwater Navigation

. . . after which the waterway winds tightly through open watermeadows to Springfield Lock, which gives access to Springfield Basin.

There has been extensive residential development by the lock, including Lockside Marina where modern apartments flank three sides of a marina, accessed by boat from the Basin by a bascule bridge (drawbridge).

Bridge over Tranquil Water 2 *– under the A138 road bridge on the approach to Springfield. A composition of modern and bygone infrastructure*
* *10-20mm lens at 18mm, f22.0, 1/20 sec, ISO 100, WB 5500*

Springfield Lock – with the residential development and marina behind the camera
* *18-50mm lens at 26mm, f22.0, 1/6 sec, ISO 100, WB 6500, ND Graduated Filter*

Springfield Basin 1
– approaching the end of the canal at Chelmsford
• *18-50mm lens at 40mm, f22.0, 0.3 sec, ISO 100, WB 6200, centre spot diffuser filter*

ARTISTIC EFFECTS

In the days before digital photography special effects were performed by the use of filters placed in front of the lens such as the image on the left. Also some amazing effects were produced in the dark room.

With digital there are powerful software programs used to process the images from the camera. Bolted on to these programs are all the tools you need to emulate the old darkroom techniques, and much more, as shown below.

Springfield Basin 2 *– nearing the end of the canal with digitally applied art effect*
• *18-50mm lens at 40mm, f22.0, 1/10 sec, ISO 100*

The Chelmer and Blackwater Navigation

Hence we arrive at Springfied Basin which was derelict for many years until restoration and reopening by IWA in 1993.

Now it is home to a floating barge-based office enterprise, Springfield Barges. Moorings are provided at the head of the Basin, within an easy walk of the town centre, overlooked by a smart restaurant.

From here the city of **Chelmsford** can be explored.

Springfield Basin 3 – the view from the end of the
Blackwater and Chelmer Canal at Chelmsford
• 18-50mm lens at 23mm, f22.0, 1/4 sec,
ISO100, WB 5500

Maldon District

. . . dominated by the district's 60 miles of coastline . . .

Maldon District covers an area of over 36,000 hectares in the county. The landscape and character are dominated by the district's 60 miles of coastline that includes the estuaries of the rivers *Blackwater* and *Crouch*.

The district is rich in wildlife and has many natural attractions including saltwater marshes, farmland and a variety of charming villages.

The main town of Maldon itself is thought to be among the oldest recorded towns in the county and records show that settlements in the area go back to Saxon times.

Thames Barge – *silhouetted against the early morning sun*
- *18-55mm lens at 27mm, f22.0, 1/320 sec, ISO 100, WB 4850, ND Graduated Filter*

Oyster Smack – *silhouetted against the early morning sun*
- *18-55mm lens at 21mm, f22.0, 1/60 sec, ISO 400, WB 4500, ND Graduated Filter*

Maldon Boats – Thames Barge 'Thistle' setting off from the quayside.

Thames barges played a vitally important part in the life of East Essex by making use of the many creeks and inlets to collect the produce from the farms and take it to the markets.

Barges were very flat bottomed wooden craft to ensure that they could navigate the many shallow inlets by sailing in as little as 3 feet of water and then lie on the mud without tilting.

They were sized 80 feet long by 20 feet wide to ensure that they could carry a large load.

A crew of three, a skipper, mate (related) and the family dog, sailed these barges which was said to be physically hard work as they sailed in all weathers.

The distinctive sails have a large topsail, mainsail and foresail on two masts. The rusty red colour is due to the waterproofing traditionally used on the sails.

• 55-200mm lens at 85mm, F5.6,
  1/800 sec, ISO 200, WB 6700

Maldon District

Moored Barges – Thames Barges moored for the evening at Maldon prior to the Race early the following morning from Osea Island
• 18-50mm lens at 50mm, F22.0, 1/10 sec, ISO 100, WB 8000

Maldon's sailing barges were a major industry all along the east coast of the UK with the area supplying much of London with agricultural products.

Nowadays these famous Thames Barges are mainly used for charter or tourist purposes. One of the eight Thames Barge Races, with up to twenty barges sailing together, is held here annually.

Thames Barge Triptych – participating in the annual Race from Osea Island at Maldon to Mersea Island
• 150-500mm lens at 150mm, F22.0, 1/13 sec, ISO 100, WB 8000

Thames Barges *– participating in the
annual Race from
Osea Island at Maldon to Mersea island*
*• 150-500mm lens at 174mm, F22.0,
1/20 sec, ISO 100, WB 8000*

USING VIGNETTES

*Unwanted vignetting occurs when using too many filters or using a cheaper wide angle
lens at full aperture creating a darkening in the corners.*

*However, deliberate vignetting (darkening the corners) can be used to enhance an image
by placing greater emphasis on important elements within the picture. I achieve this
while the image is still in the RAW format. Navigate to the Lens Corrections icon and
drag the Amount slider to the left for the desired effect and then drag the Midpoint slider
to the left which increases the size of the vignetting and creates a soft pleasing effect.*

Black Headed Gull *– (Larus ridibundus)
on the quayside at Maldon*
*• 55-200mm lens at 200mm, F5.6,
1/1250 sec, ISO 200, WB 7500*

Maldon District

Masts & Rigging – of the moored Thames Barges at the quayside
• 55-200mm lens at 160mm F5.6, 1/800 sec, ISO 200, WB 6000

The Hythe quayside not only harbours the barges but also accommodates the Maldon Day of Dance as one of the venues on the weekend circuit.

Usually held on a weekend in June and hosted by the three resident morris sides: Maldon Greenjackets (cotswold morris), Dark Horse (border morris) and Alive and Kicking (north west clog). Normally three visiting sides also participate in the weekends merrymaking.

Records show that morris was danced in Maldon as far back as 1540, where the Chamberlains' accounts refer to payments for bells, minstrels and morris dancers.

Chelmsford Morris Taking Five – (both mens and ladies sides) relaxing on the quayside
• 55-200mm lens at 80mm, F5.6, 1/1000 sec, ISO 200, WB 6700

Gong Scourers – from Crayford one of the
visiting sides at the Maldon Day of Morris
• 55-200mm lens at 200mm, F5.6,
1/400 sec, ISO 200, WB 6700

Taking it Easy – spectator soaking
up the vibes and ambience on the
quayside at the Maldon Day of Morris
• 55-200mm lens at 200mm,
F5.6, 1/500 sec, ISO 200, WB 6000

Chelmsford Morris – one of the sides
at the Maldon Day of Morris
• 55-200mm lens at 165mm,
F5.6, 1/800 sec, ISO 200, WB 6700

Danbury and Lakes

. . . surrounded by attractive open countryside and woodland . . .

OK, so now we are going to go inland a few miles to **Danbury** which is a large village near Chelmsford. Situated on a hill at 365 feet above sea level, it occupies one of the highest points in Essex.

The village is surrounded by attractive open countryside and woodland, much of it protected by the Essex Wildlife Trust and the National Trust.

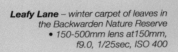

Leafy Lane – *winter carpet of leaves in the Backwarden Nature Reserve*
* *150-500mm lens at150mm, f9.0, 1/25sec, ISO 400*

Leaning Trees – *woodland track at Backwarden Nature Reserve*
* *10-20mm lens at 20mm, f16, 1.0 sec, ISO 100, WB 5500*

Winter Warmth –
an enchanting lane on
National Trust woodland
in December
• 150-500mm lens at
500mm, f9.0, 1/200sec,
ISO 400

Danbury and Lakes

Brown Rat – *(Rattus norvegicus) foraging for food at Danbury Lakes*
• *150-500mm lens at 500mm,*
f7.1, 1/250sec, ISO 400

Spring Bullrushes – *early*
spring at Danbury Lakes
• *55-200mm lens at100mm,*
f5.6, 1/640sec, ISO 400

Autumn Golds 1 – *autumn morning at Danbury Lakes*
• *18-55mm lens at 25mm, f22.0, 1/20 sec,*
ISO 100, WB 5500

Autumn Golds 2 – autumnal afternoon at Danbury Lakes
• 18-55mm lens at 22mm, f11.0, 1/10 sec,
ISO 400, WB 5500

Danbury Lakes is set within the grounds of Danbury Palace and offers a blend of ornamental gardens and lakes, together with adjoining woodland and meadows.

Late spring brings spectacular displays of rhododendrons in the gardens. It is also a good time to see the new families of ducklings and other birds around the lakes.

Most of the pictures here were taken in Autumn with resplendent golds, russet reds and greens.

Autumn Golds and Reds – autumn morning at Danbury Lakes
• 18-55mm lens at 18mm, f22.0, 1/30 sec, ISO 100, WB 7500

Danbury and Lakes

In the summer the meadows are filled with wildflowers and grasses which in turn attract a wealth of butterflies and insects.

In contrast, the paths into the woodland offer respite from the sun in the dappled shade of ancient oaks, hornbeams and sweet chestnuts.

Bottoms Up *– pair of Mallard Ducks (Anas platyhynchos) foraging for food in Danbury Lakes. Possibly forerunners to the synchronised swimming event at the Olympics*
• 150-500mm lens at 439mm, f6.3, 1/320 sec, ISO 400, WB 4600

Autumn Red *– autumnal afternoon at Danbury Lakes*
• 18-55mm lens at 18mm, f22.0, 1.6 secs, ISO 100, WB 5150

Moorhen – *(Gallinula chloropus) strutting up a river bank in Danbury Lakes*
• *150-500mm lens at 500mm, f9.0, 1/100 sec, ISO 400, WB 4500*

Old Man of the Woods –
ancient woodland in Danbury Lakes
• *10-20mm lens at 10mm, f9.0, 1/250 sec, ISO 400, WB 5500*

Autumn Golds 3 – *autumnal afternoon at Danbury Lakes*
• *18-55mm lens at 18mm, f11.0, 0.5 sec, ISO 100, WB 5500*

Woodham Ferrers District

. . . Originally a hermitage during the reign of Henry II . . .

Woodham Ferrers is a small village about 7.5 miles southeast of Chelmsford, located between South Woodham Ferrers and Bicknacre, with Rettendon to the west and Stow Maries to the east.

Originally a hermitage during the reign of Henry II the name Woodham was adopted in 1175 when it became a priory, including 60 acres of forest stretching towards Danbury.

St Mary's Church, situated at the south end of the village, was recorded in the Domesday Book in 1066.

Summer Sunset – a field of rapeseed lunar lit at sunset
• 18-50mm lens at 50mm, f16.0, 1/15 sec, ISO 100, WB 5150, ND Graduated Filter

Winter Blue – December snow with a clear blue sky
• 18-50mm lens at 50mm, f22.0, 1/200 sec, ISO 400, WB 5000

Yellow Field 1 – a field of rape (Rapum turnip) from Woodham Ferrers looking towards Crouch Vale with South Woodham Ferrers in the distance
• *18-50mm lens at 40mm, f22.0, 1/10 sec, ISO 100, WB 6700*

Yellow Field 2 – a field of rape (Rapum turnip) cultivated for its seeds (rapeseed), which yield a useful oil (rape oil), and as a fodder plant.
• *10-20mm lens at 20mm, f22.0, 1/15 sec, ISO 100, WB 5000*

The area is typically Essex rural countryside with early summer fields of rapeseed and late summer fields of wheat and barley. Undulating farmland is interspersed with small woodlands, the views interrupted by mighty oaks and ashes presided over by the hurrying skies.

The Essex countryside is a walking paradise, whether you want to cross the country from south-west to north-east using the well-known Essex Way or other smaller countryside trails or coastal walks.

Woodham Ferrers District

Straw Bales – traditionally a waste product which farmers do not till under the soil, but sell as animal bedding or landscape supply due to their durable nature. In many areas of the country, it is also burned, causing severe air quality problems.

Baled straw from wheat, oats, barley, rye, rice and others is used for construction purposes in walls covered by earthen, lime or cement stucco (fine plaster).

It is important to know that straw is the dry plant material or stalk left in the field after a plant has matured, been harvested for seed, and is therefore, no longer alive.

Hay Bales – in comparison are made from short species of livestock feed grass that is green and, therefore, still alive.

__Straw Bales 1__ – harvesting the grain late summer in Rettendon
• 55-200mm lens at 170mm, f22.0, 1/50 sec, ISO 100, WB 5150

__All Saints Church Rettendon__ – sitting on a hill the church dates back to Norman times but is mainly of the perpendicular style. It has a large tower which is a landmark for miles around
• 55-200mm lens at 95mm, f22.0, 1/10 sec, ISO 100, WB 5150, ND Graduated Filter

Straw Bales 2 *– harvesting the grain late summer in Stow Maries*
* *55-200mm lens at 200mm, f22.0, 1/13 sec, ISO 100, WB 5150*

Straw Bales 3 *– harvesting the grain late summer in Stow Maries*
* *55-200mm lens at 80mm, f22.0, 1/13 sec, ISO 100, WB 5150*

Blackwater Estuary

. . . The estuary is well known for its sailing qualities . . .

So, enough of the inland diversion, back to Maldon together with the **Blackwater Estuary** and a personal indulgence of mine. Namely, the very essence of the Essex coastline and waterways (that is if you haven't guessed already) – boats!

These next few pages are the result of a thoroughly enjoyable day trip aboard the Thames sailing barge *Thistle* (see page 123) on one of the scheduled barge races that occur during the summer months around the Essex coast – *enjoy!*

Departure from Maldon Quayside – *looking back to the town of Maldon as we depart from the quay*
• *55 to 200mm lens at 125mm, f9, 1/320 sec, ISO 100, WB 5150*

Thames Barges – *'Reminder'* and *'Decima'*.
 • *55 to 200mm lens at 85mm, f9, 1/320 sec, ISO 100, WB 5150*

Thames Barge – *'Repertor'*.
 • *55 to 200mm lens at 180mm, f9, 1/125 sec, ISO 100, WB 5150*

Although it is one of the largest estuaries on the east coast of England the river is a comparatively small stream.
Its source is on the edge of Debden Airfied in North West Essex where it is known as the *River Pant*.

It is not until it reaches Braintree that it becomes the *River Blackwater*, continuing on past the weir at Beeleigh to reach the estuary and the sea.

The estuary is well known for its sailing qualities and there are numerous anchoring areas both onshore and offshore with the mud banks offering a suitable overnight base for yachtsmen. Sailing Clubs with good facilities are found at regular intervals along its banks.

The estuary is also a major stopping point on the migration routes of birds flying between the Arctic and Africa.

The sea walls, shell banks, salt marshes and inland drainage dykes provide a superb habitat for many marine and land based wildlife and as such most of the area has been declared a Site of Special Scientific Interest (SSSI).

Racing Thames Barges –
• 55 to 200mm lens, aperture priority f9, ISO 100, WB5150

In 1907 there were 2090 Thames sailing barges registered. After the First World War barge building was in decline although many barges were built particularly in steel. The last wooden barge built was the 'Cabby' launched in 1928 and the final steel barge 'Blue Mermaid' was launched from Mistley, here in Essex, in 1930.

Today there are around two dozen or so that are in regular sea-going use, either privately or as charter vessels. Many of these also compete in the Barge Match Races each year as seen here. There are also some twenty or so that are under restoration, many of these are very long term projects and some may never be completed due to the prohibitive cost of such work.

With their large cargo hatches, cheap construction and the unique rig (which stowed all the sails out of the way of loading and unloading cargo) Thames sailing barges were perfect for the job they were designed to do. With a shallow draft they were able to sail into the coastal creeks in two to three feet of water. No other craft in Europe ever attained the numbers that the Thames sailing barge did.

They carried a variety of cargoes. Delivering the stone bricks and timber from which modern London was built in the 19th century. The city depended then upon thousands of horses for its transport and it was the Thames sailing barge that delivered the huge requirements of hay and straw from the farms of Essex, Suffolk and Kent.

Essex and Maldon in particular evolved a special variety of barge called the "stackie" designed to be shallow and wide for sailing with a haystack on deck.

Barges also carried timber, stone, sand, cement, ballast, bricks, oilcake, oil and plastics. In fact everything that small ships can carry was carried regularly by the barges.

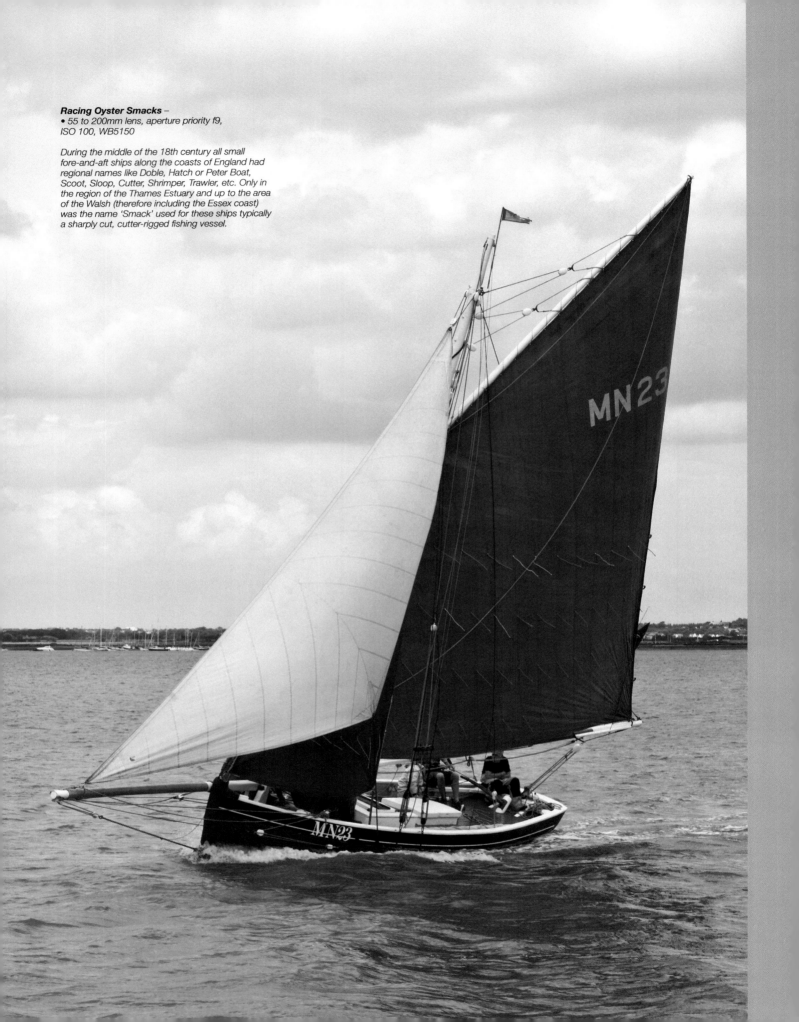

Racing Oyster Smacks –
• 55 to 200mm lens, aperture priority f9,
ISO 100, WB5150

*During the middle of the 18th century all small
fore-and-aft ships along the coasts of England had
regional names like Doble, Hatch or Peter Boat,
Scoot, Sloop, Cutter, Shrimper, Trawler, etc. Only in
the region of the Thames Estuary and up to the area
of the Walsh (therefore including the Essex coast)
was the name 'Smack' used for these ships typically
a sharply cut, cutter-rigged fishing vessel.*

Racing Sails – *from numerous marinas mingling with the 2012 Olympic squad?*
• 55 to 200mm lens, aperture priority f9, ISO 100, WB5150

Racing Catamarans – *notice the GBR on the sails - possibly training for the 2012 Olympic Games?*
• 55 to 200mm lens, aperture priority f9, ISO 100, WB5150

Blackwater Estuary

WATERCOLOUR EFFECTS FROM 'OUT OF FOCUS' PHOTOGRAPHS

Out of focus due to slow shutter speed! Well all is not lost. By using a few adjustments in a post digital darkroom package you can turn the image into many artistic lookalikes.

In this case I gave the image a watercolour appearance by using selective adjustments on colour, contrast, layer blends, blurs, sharpening and noise.

Blackwater Pillbox *– on the north bank of the estuary. No one is quite sure how many were built in the first place, but research suggests it was about 28,000, of which about 6,500 still survive. The thickness of the walls and roofs varied from a foot to 3ft 6ins.*
* 55 to 200mm lens at 200mm, f9, 1/100 sec, ISO 100, WB 5150

Windsurfing – along with kitesurfing is very popular along the coast
• 55 to 200mm lens at 200mm, f9, 1/250 sec, ISO 100, WB 5150

Bradwell Chapel and Sales Point

. . . overlooking the entrance to the Blackwater Estuary for over 1300 years . . .

A long stretch of Roman road leads east from the village of **Bradwell**, becoming a track and then a path. At the end of this path, where the sea meets the sky, is the oldest church in England.

St Peter's on the Wall was built by St Cedd from the ruins of Othona, a Roman fort, which probably included part of a wall.

He landed in 653, on his mission from Lindisfarne to lighten the Dark Ages of the heathen East Angles. The church has stood overlooking the entrance to the *Blackwater Estuary* for over 1300 years.

Inside – simplicity, beauty and peace. A feeling of spirituality and history. Consecrated in 654 AD, this is the earliest Cathedral in the country of which so much remains.
• 10 to 20mm lens at 14mm, f22, 1/40 sec, ISO 100, WB 4000

The Approach – made by approximately 1000 pilgrims per month. This same road has been used by visitors to the Chapel for over 1300 years. Before that it was used by the Romans.
• 18 to 50mm lens at 23mm, f11, 1/200 sec, ISO 100, WB 5150, ND Graduated Filter

Ancient History of Essex
ST. PETER'S ON THE WALL

1300 years ago there were people working to spread the Christian faith from which many monasteries and Christian centres were established.

Lindisfarne on the north-east coast was one such establishment and is where Anglo-Saxon boys such as Cedd and his brothers Caelin, Cynebil and Chad learnt to read and write in Latin, to teach the Christian faith and train to become priests and missionaries. All four brothers were ordained as priests and two of them, Cedd and Chad, later became bishops.

Cedd's first mission was to go to the Midlands (then called Mercia) at the request of its king who wanted his people to become Christians. Cedd was so successful that King Sigbert of the East Saxons (Essex) asked for a similar mission.

So in 653 Cedd sailed down the coast from Lindisfarne and landed at Bradwell. Here he found the ruins of an old deserted Roman fort called Othona. He probably first built a small wooden church but as there was so much stone from the fort he soon realised that it would provide a much more permanent building. So he replaced it the next year with the chapel we see today! In fact he built his monastery where the gatehouse of the fort had been. So it was built on the wall of the fort, hence the name, Saint Peter-on-the-Wall.

St. Peter's on the Wall – viewed from the sea wall with a derelict bird hide overlooking the nature reserve and cockle spit
- *10 to 20mm lens at 15mm, f16, 1/40 sec, ISO 100, WB 5150, ND Graduated Filter*

Cedd's mission to the East Saxons was so successful that the same year he was recalled to Lindisfarne and made Bishop of the East Saxons. His simple monastery at Bradwell would, like the one at Lindisfarne, have been at the same time a church, a community of both men and women, a hospital, a library, a school, an arts centre, a farm, a guest house and a mission base. From here he established other Christian centres in Essex at Mersea, Tilbury, Prittlewell and Upminster.

Outside – the fabric of the old ruined fort of Othona
- *70mm lens, f22, 1/25 sec, ISO 100, WB 4700*

Bradwell Chapel and Sales Point

Walking north along the sea wall we come to the northern most end of Bradwell Cockle Spit at **Sales Point**, situated at the mouth of the *Blackwater Estuary*, with sunken barges in the background protecting the saltmarsh and seawall from erosion.

The Spit consists of 252 acres of saltmarsh and shingle with 30 acres of shell bank. It is a haven for wading birds and other wildlife.

Sunken Barges 1 – close up view at low tide
• 18 to 50mm lens at 24mm, f11, 1/160 sec, ISO 100, WB 5150, ND Graduated Filter

Cockle Spit – The northernmost end at Sales Point with sunken barges in the background protecting the site.
• 10 to 20mm lens at 18mm, f22, 1/10 sec, ISO 100, WB 5050

The spit has been built up by tidal currents and is mainly cockle and oyster shells. During bad weather and strong tides, the spit moves and in the past two decades, up to 100 metres of saltmarsh has been lost to erosion.

Sunken Barges 1 – *viewed from the sea wall at low tide*
• *10 to 20mm lens at 14mm, f11, 1/320 sec,*
ISO 200, WB 5150, ND Graduated Filter

Nissen Hut – *by Sales Point, invented and built as housing for troops*
in WW1 (The Great War).
• *70mm lens, f9, 1/125 sec, ISO 100, WB5250*

St Peter's on the Wall – *viewed from the sea wall at Sales Point*
• *70 to 300mm lens at 300mm, f16, 1/80 sec,*
ISO 100, WB5150, ND Graduated Filter

The Restoration of St. Peter's on the Wall

In 1920 the owner of the land on which St Peter's stands returned the Chapel to the Chelmsford Diocese and it was restored and then re-consecrated by the Bishop of Chelmsford on 22 June 1920.

The altar was consecrated in 1980 by the Anglican Bishop of Chelmsford and the Roman Catholic Bishop of Brentwood. In the altar are stones donated by the religious communities at Iona, Lindisfarne and Lastingham. The original Nave still remains as it was in the time of St Cedd.

Visitors often remark on the calmness felt in the building and the sense of awe when considering its history.

The River Crouch Navigation

. . . noted for its sailing qualities, numerous anchoring areas both onshore and offshore . . .

The source of the *River Crouch* is in **Little Burstead** rising from a series of springs into ponds, set in beautiful wooded glades. Known locally as The Wilderness it is set in the middle of a golf course and, as a golfer myself I had a bonus day by leaving this beautiful spot with a bag of lost golf balls!

It then runs parallel with the A176 for approximately 1.3 miles (2.1 km) and then follows a generally easterly course, passing near to, or through, Crays Hill, Ramsden Bellhouse, Wickford (where it flows along a concrete culvert) and on to Runwell.

The Wilderness 1 *– the source of the River Crouch which rises from a series of springs into ponds*
• *10-20mm lens at 20mm, f22.0, 2.0 sec, ISO 100, WB 5150*

The Wilderness 2
• *10-20mm lens at 20mm, f22.0, 1/3 sec, ISO 100, WB 5150*

The Wilderness 3
• 10-20mm lens at 18mm, f22.0, 1.0 sec, ISO 100, WB 5150

The River Crouch Navigation

The river then meanders on to **Battlesbridge**, the head of the tidal *River Crouch Navigation*, flowing in a wide estuary for 17.5 miles until it reaches the North Sea at Holliwell Point (north bank) and Foulness Point (south bank).

The *Navigation* is noted for its sailing qualities, numerous anchoring areas both onshore and offshore and the mud banks offering a suitable overnight base for yachtsmen.

Antique Centre – dominated by the old mill and where the original wooden bridge has been replaced by an iron bridge
• *18-50mm lens at 20mm, f22, 1/40 sec, ISO 100, WB 5150*

Hamlet of Battlesbridge – viewed from the bridge, a part of this picturesque hamlet
• *18-50mm lens at 20mm, f11.0, 5 sec, ISO 100, WB 5150,*
    ND Graduated + ND Filter

Battlesbridge, a picturesque hamlet, is a conservation area jointly designated by Chelmsford Borough Council (north side of the river) and Rochford District Council (south side) in 1992.

Now an iron bridge (see opposite) the original wooden bridge over the tidal part of the river was maintained in the early days by the Bataille family, hence the name of the hamlet.

Today it is home to a number of antiques centres, one of which used to be a mill (see opposite). Classic car and motorbike shows are held here each year.

The tidal gates (see below) were replaced in 2008, weighing 19 tonnes, and constructed using pine beams and metal sluices.

Tidal Gates – *the tidal gates where river meets estuary*
• *18-50mm lens at 35mm, f11.0, 3.2 secs, ISO 100, WB 5150, ND Graduated + ND Filter*

The River Crouch Navigation

The navigation then flows between **South Woodham Ferrers** (north side of the river) and **Hullbridge** (south side).

Development of the modern town of South Woodham Ferrers began in the 1960's. Rooftop horizons mimic wharves, sail lofts and farm cottages.

Nearby Marsh Farm Country Park is a working farm with visitors centre and lunch area. The park is also a good starting point for some excellent riverside walks.

River Crouch Calm – near the sailing club at South Woodham Ferrers
• 150-500mm lens at 150mm, f9.0, 1/3200sec, ISO 400, WB 5150

March Sunrise – on the River Crouch one cold morning in March
• 18-50mm lens at 24mm, f22.0, 1/4sec, ISO 100, WB 5500, ND Graduated Filter

Woodham Boats – *view of the River Crouch taken from South Woodham Ferrers Sailing Club*
• *55-200mm lens at 65mm, f22.0, 1/5sec, ISO 100, WB 5500, ND Graduated Filter*

Black Headed Gulls 1 – *(Larus ridibundus) a pair of gulls in winter plumage on the tidal River Crouch*
• *150-500mm lens at 313mm, f11.0, 1/80sec, ISO 200, WB 5500*

The village of Hullbridge takes its name from a Roman bridge that once crossed the river at this point.

With over two miles of frontage to this part of the river the parish is well used by the local boating fraternity with hundreds of moorings and three yacht clubs.

Along the river bank there are remains of evaporation pans where salt was made in medieval times.

The River Crouch Navigation

Herring Gull – (*Larus argentatus*) on the
River Crouch at South Woodham Ferrers
• 150-500mm lens at 500mm, f9.0,
1/80sec, ISO 400, WB 5350

The *Crouch and Roach Estuaries* are of international importance for wildlife. The relatively mild climate and abundance of food attract internationally important numbers of wild fowl and waders during the winter months.

Nearly twenty-five thousand water birds visit the estuary each year including nationally important numbers of shelduck, shoveler and black-tailed godwit and internationally important numbers of dark bellied Brent Geese.

Mute Swans – (*Cygnus olor*) on the River Crouch near South Woodham Ferrers
• 150-500mm lens at 500mm, f9.0, 1/400sec, ISO 400, WB 4750

The Essex Coast provides over-wintering for around one fifth of the world population of dark bellied Brent Geese with an average peak of just over 6 thousand birds (about 2.5% of the world population) congregating around the *Crouch and Roach Estuaries*.

Black Headed Gull and Lapwings – *Black Headed Gull in winter plumage (Larus ridibundus) accompanied by Lapwings (Vanellus vanellus) on the tidal River Crouch near South Woodham Ferrers*
• 150-500mm lens at 500mm, f9.0, 1/100sec, ISO 400, WB 5000

Turnstones – *(Arenaria interpres) on shoreline of the tidal River Crouch near South Woodham Ferrers*
• 150-500mm lens at 500mm, f9.0, 1/60sec, ISO 400, WB 5850

The River Crouch Navigation

The Drama Queen – wrecked barge on the eastern tidal creek near
South Woodham Ferrers early one icy cold morning in February at low tide
• 18-50mm lens at 50mm, f22.0, 0.6 secs, ISO 100, WB 5300, ND Grad Filter

The *Navigation* is famous for its mud and salt
marshes which are interwoven by a multitude of
little creeks. Once the haunt of smugglers, this
whole area is now a Site of Special Scientific Interest
(SSSI) and is considered one of the few true
wildernesses left in England.

Essex is often brushed aside as flat and boring but
this is probably because some of the best parts are
difficult to get to and more easily reached by water
than by land.

South Woodham Ferrers is sited between two tidal
creeks to the east and west and the *River Crouch* to
the south.

Creek View – the curving eastern tidal creek leading to Clements Green Creek
at sunrise and at low tide
• 18-50mm lens at 24mm, f22.0, 1.6 secs, ISO 100, WB 6700, ND Grad Filter

The Drama Queen 2 – wrecked barge on the eastern tidal creek near South Woodham Ferrers early one icy cold morning in February at low tide
• 18-50mm lens at 18mm, f22.0, 1.6 secs, ISO 100, WB 7500, ND Graduated Filter

The navigation has eight yacht clubs and three harbours at Fambridge, Wallasea and Burnham.

Water skiing is unlawful on most of the *River Crouch* and is totally banned on the *River Roach*. Areas at Clements Green Creek (Woodham Ferrers), Hayes Farm and Brandy Hole have been designated as suitable for water skiing.

Care needs to be taken in the navigation as the fast flowing water, where shallows are frequent, can prove hazardous especially in adverse wind conditions.

The Drama Queen 3 – a close view of the barge. The only film image in the book taken with my old Olympus OM10, consequently, I did not write down the exposure details

The River Crouch Navigation

Grey Heron – *(Ardea cinerea) resting on tidal saltmarsh at Blue House Farm Nature Reserve, Fambridge*
• *150-500mm lens at 500mm, f6.3, 1/500 secs, ISO 100, WB 4850*

The river now flows on into the Dengie peninsular reaching between the two rural parishes of **North Fambridge** and **South Fambridge**.

Between the two parishes there used to be a ferry which operated for hundreds of years across the river until the 1940's or 50's.

North Fambridge has a jetty for moorings, boatyard, and a nearby public house with restaurant and accommodation.

Boats at Fambridge 1 – *moored boats off the jetty on the River Crouch at Fambridge*
• *55-200mm lens at 55mm, f22.0, 1/40 sec, ISO 200, WB 4750, ND Graduated Filter*

Boats at Fambridge 2 – *Moored boats on the River Crouch at Fambridge*
• 18-55mm lens at 39mm, f22.0, 1/200 secs, ISO 200, WB 7500, ND Graduated Filter

Gatekeeper Butterfly –
*(Pyronia tithonas) resting on
Saw-wort (Serratula tinctoria)
taken from inside one of the
bird hides on Blue House Farm
Nature Reserve, Fambridge
• 150-500mm lens at 500mm,
f11.0, 1/160 sec, ISO 200,
WB 5300*

The River Crouch Navigation

North Fambridge is also surrounded by farmland and is adjacent to the Essex Wildlife Trust site of Blue House Farm. The area being subject to a wide range of international, national and regional designations.

Waterside Plants – a study of various waterside plants taken from inside a bird hide at Blue House Farm Nature Reserve
• 55-200mm lens at 200mm, f5.6, 1/640 sec, ISO 400, WB 4600

Boats at Fambridge 3 – evening view of moored boats at Fambridge
• 18-50mm lens at 31mm, f22, 5 sec, ISO 100, WB 6000, ND Graduated + ND filter

Desolate Abode – *riverside dwelling at Fambridge*
• *18-50mm lens at 35mm, f22, 6 sec, ISO 100,*
WB 6900, ND Graduated + ND filter

. . . After the battle, England became and always remained a unified kingdom and a single nation . . .

Ancient History of Essex

THE BATTLE OF ASHINGDON

Source: Ashingdon Parish Council
www.essexinfo.net/ashingdonparish/history

Well worth a mention folks as this main event occurred just up the road from South Fambridge!

Ashingdon has been a village for more than one thousand years. It was called 'Nesenduna' in the 900s and 1000s and it has had many spellings over the centuries.

The village appears in the Domesday Book produced for King William The Conqueror in 1085 and 1086.

Back to The Dark Ages: for hundreds of years after the Romans left in about 420AD, England was occupied, divided and ruled by various groups who could be called invaders, settlers or newcomers from neighbouring countries.

They included the Saxons and Angles, the two largest early settlers. Then came the Danes and Vikings who mainly settled in the Northeast, the Northwest and in Eastern England.

By the 700s to 800s, the East Saxons and the East Angles had been taken over by the Mercians and their kingdoms became part of the Mercians' East Anglia.

By the 900s, The Danes had taken over East Anglia, East Mercia and Eastern and Northeastern England. The Danes also took over most of Norway and so their Viking territory in the Northwest and Northeast of England was combined with the Dane's territory in the East of England.

There were various kings of the two parts of the now divided England. One part was The Danelaw which included the east, East Midlands, northeast and northwest and was ruled by Danish kings. The other part was south and west of England from Kent, along the South Coast, the west and West Midlands which was ruled by The Saxons and called Wessex and West Mercia.

Some of those kings, both Danish and Saxon, claimed to be the king of all England.

The 'Danelaw' boundary between Danish England and Saxon England was along geographical and historic features: from The North Sea, along The River Thames to Stratford, along The River Lea, north of Luton to Watling Street, the old Roman Road. It followed that road past Rugby, north of Birmingham, through the Midland Gap, along The River Dee to the Irish Sea.

At the time of the struggles between the two powers there had been many disputes, claims, conflicts and battles. The most decisive was to take place at Ashingdon.

King Canute (the Danish king) had fought in several conflicts against King Edmund for control of London and the South of England.

He withdrew to his stronghold of East Anglia in his Danelaw (which included Essex). He sailed into the River Crouch towards Ashingdon and set up his camp at Canewdon.

King Edmund Ironside (the Saxon king) heard of this and he came with his army to Ashingdon which became his base camp.

The claim for the control and rule of England was settled by this battle which took place here at Ashingdon in 1016 AD.

Canute fought Edmund and won the battle and, soon after, secured the Kingdom of all England. He established full control of not only the Danelaw, but also the Saxon Kingdom. After the battle, the Saxon area was administered by the Saxons on behalf of the Danish King of (all) England. The Danelaw was ruled directly by the King.

Then when Edmund died, the Danes ruled solely from 1016 until 1042. After that, the Saxons ruled solely until 1066, when Duke William 'The Conqueror' and the Normans invaded.

Prior to The Battle of Ashingdon, what we call England had almost always been divided. It was two kingdoms - the Danelaw, and Wessex and West Mercia.

After the battle, England became and always remained a unified kingdom and a single nation, as it is to this day.

Without the Battle of Ashingdon, King Harold would have had a smaller kingdom to defend more easily and William The Conqueror may not have won at Hastings. Or, William would have gained only Wessex and West Mercia, resulting in the Danelaw remaining part of the Kingdom of Denmark.

After the battle, Canute had a Church built in 1020 to honour the defeated but pious and devout King Edmund and all those who died in the battle. He returned to Ashingdon to open the church that he had built which was known then as Ashingdon Minster - now known as St. Andrew's Church. Most of what still stands is mediaeval and some of it may be based on the original Minster.

The first priest at Ashingdon Minster was a young man named Stigand. By 1052 Stigand had progressed within the clergy to the highest position within the church to become The Archbishop of Canterbury, a position that he held until 1070.

In that capacity, he crowned King Harold as King of England in early 1066 (as portrayed in the famous Bayeux Tapestry) and then on Christmas Day, 25th December 1066, it is believed he crowned King William The Conqueror as King of England.

THE BAYEUX TAPESTRY

The Bayeux Tapestry is preserved and displayed in Bayeux, in Normandy, France. Nothing is known for certain about the tapestry's origins.

Archbishop Stigand was Ashingdon's first priest when he was a young man at the opening of the new church in 1020. He is depicted on the Bayeux Tapestry standing beside King Harold after crowning him.
The inscription on the Tapestry in Latin (with missing letters in brackets) says :

"hIC DEDERRUNT: HAROLDO: CORONA(m): REGIS".

"hIC RESIDET: hAROLD REX: ANGLORUM:
STIGANT ARChIEP(iscopu)S".

Meaning in English - with missing words in brackets :

"Here they give up (the) royal crown (to) Harold".

"Here sits Harold (the) King of England.

"Stigant Archb(ishop)".

The River Crouch Navigation

We now come to Bridgemarsh Marina situated on Althorne Creek. On the south side of the creek is Bridgemarsh Island which separates the creek from the *Crouch*.

The Marina is one to two miles from the village of **Althorne**, home to Phillip Scott Burge, a top British fighter ace in World War One with 11 enemy kills between March and July 1918.

The name Althorne has an unusual meaning in Old English, translating as '(place at) the burnt thorn-tree'.

Dinghies and Rowing Boats – *on the south shore*
• *10-20mm lens at 10mm, f16, 1/10 sec, ISO 100, WB 5900*

Old Mooring – *a view of the marina with Bridgemarsh Island in the background*
• *10-20mm lens at 10mm, f16, 1/13 sec, ISO 100, WB 4800*

Shoreside Gangway – a view of the marina with Bridgemarsh Island in the background
• 10-20mm lens at 10mm, f9.0, 1/50 sec, ISO 100, WB 5200

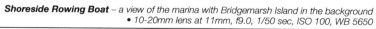

Shoreside Rowing Boat – a view of the marina with Bridgemarsh Island in the background
• 10-20mm lens at 11mm, f9.0, 1/50 sec, ISO 100, WB 5650

EVENING LIGHT

These images were taken on an early summers evening with the low sun giving a feeling of warmth as you can see.

The colour of a scene is constantly changing depending on the ambient lighting conditions at the time.

Generally, as the day goes on the colour cast warms up and by late afternoon/late evening you should expect to have a warm golden glow.

By sunset most landscapes assume a distinctive orange/red glow.

The River Crouch Navigation

It is quite likely that the name 'Crouch' is derived from 'Creek' or 'Crick' after the two places called **Creeksea** either side of the river which had an old ferry between them.

The Creeksea near Burnham on the north shore of the *Crouch* is a tiny hamlet with two manors, a few houses and farms and a ferry jetty.

Creeksea on the south shore is a small part of Wallasea Island which is now dominated by the Essex Marina and Baltic Wharf. There is a small boat that will ferry people to and from Burnham as required.

__Baltic Wharf__ – on Wallasea Island viewed from the jetty at north shore Creeksea
* *18-50mm lens at 50mm, f18, 1/50 sec, ISO 100, WB 5150*

__Creaksea Boatyard__ – taken from the ferry jetty with Bridgemarsh Marina in the distance
* *18-50mm lens at 29mm, f18, 1/80 sec, ISO 100, WB 5150*

Essex Marina – on Wallasea island viewed from
near the jetty at north shore Creeksea
• 18-50mm lens at 50mm, f18,
1/50 sec, ISO 100, WB 5150

Authors Note:

I found this stretch of the river from Creeksea to Burnham on the north shore a most pleasant and rewarding walk.

There were fellow walkers (with or without family), joggers and mountain bikers also enjoying the picturesque surroundings.

See following page.

Burning Field – field burning in spring aleady?
• 18-50mm lens at 40mm, f18, 1/100 sec,
ISO 100, WB 5150

Sea Wall 1 – *view between Creeksea and Burnham at low tide*
• *18-50mm lens at 46mm, f18, 1/80 sec, ISO 100, WB 5150*

Sea Wall 2 – *view between Creeksea and Burnham at low tide*
• *18-50mm lens at 23mm, f18, 1/60 sec, ISO 100, WB 5150*

Baltic Wharf, Wallasea – *view from the sea wall between Creeksea and Burnham at high tide*
• *18-50mm lens at 35mm, f18, 1/60 sec, ISO 100, WB 5150*

Sea Wall 3 – *view from the sea wall between Creeksea and Burnham at low tide*
• *18-50mm lens at 18mm, f18, 1/50 sec, ISO 100, WB 5150*

Wallasea Island

Bounded to the north by the *River Crouch*, to the south east by the *River Roach*, and to the west by Paglesham Pool and the narrow Paglesham Creek.

500 years ago this corner of coastal Essex was drained of saltmarsh to make way for farmland and crop production, mainly wheat.

A few years ago a project to convert part of the island's farmland into mudflats and salt marsh was completed by bulldozing 300m of the sea defence wall, at the points of maximum pressure on the estuary. An area of 115 hectares was flooded, which is turning into wetland, mudflats, saline lagoons and several artificial islands. The wetlands are intended to provide winter grounds for wading birds, and ease flood problems on the *River Crouch*.

Now, the Royal Society for the Protection of Birds (RSPB) is about to recreate the salt marsh, turning the clock back by hundreds of years...

Essex Marina – view from the sea wall on Wallasea Island on the south shore of the river at low tide one evening in April with Baltic Wharf in the distance
• 18-50mm lens at 20mm, f16, 1/25 sec, ISO 100, WB 5150, ND Graduated Filter

The River Crouch Navigation

...the aim of this project is to combat the threats from climate change and coastal flooding by recreating the ancient wetland landscape of mudflats and saltmarsh, lagoons and pasture. It will also help to compensate for the loss of such tidal habitats elsewhere in England.

Once completed, this will provide a haven for a wonderful array of nationally and internationally important wildlife and a place for the local community, and those from further afield, to come and enjoy.

The reserve is planned to be in development until around 2019.

The current sea wall access along the north sea wall (or south shore of the River Crouch) is a fantastic place to come for walking, cycling, birdwatching, painting, photography or simply taking in the sea air.

New Sea Wall (above) – with the newly created saltmarsh to the left
• *10-20mm lens at 11mm, f16, 1/30 sec, ISO 100, WB 5150*

New Wetlands
– viewed to the left of the new sea wall at low tide and fenced off with Burnham in the distance across the river
• *10-20mm lens at 10mm, f16, 1/40 sec, ISO 100, WB 5150*

The River Crouch Navigation

A feature of Wallasea Island are these pontoon segments that once stretched across the river during the Second World War.

Shoreline Pontoons – *old wartime pontoon segments and new moorings with a view across to Burnham from Wallasea Island*
* *55-200mm lens at 80mm, f16, 1/80 sec, ISO 100, WB 5000*

Inland Pontoon – *part of an old floating bridge, sections of it are scattered all over this marsh*
* *18-50mm lens at 18mm, f16, 1/40 sec, ISO 100, WB5150*

The River Crouch Navigation

Back to the north side of the river and the approach to **Burnam-on-Crouch** (see pages 184-185).

__Houseboats 1__ – A portrait view taken from a man made shingle jetty early one morning at Burnham • 55-200mm lens at 55mm, f22.0, 1/60 sec, ISO 100, WB 6700, ND Graduated Filter

Estuaries Merge – *In the distance and to the right the River Roach merges from the left with the River Crouch at Wallasea Island. The foreground features one of nine surviving pillboxes along this section of the river. Known as an Essex Lozenge it was designed specifically to defend the county's sea walls. It is, basically, two pillboxes built back to back to form a single, eight-sided structure, with one half facing the sea and the other the marshes and fields.*
• *18-50mm lens at 21mm, f16, 1/160 sec, ISO 200, WB5150*

Several streams flow into the *Navigation* but the major tributary which joins with the *River Crouch* at Wallasea is the *River Roach*.

A few centuries ago, its name gradually altered based on the name of the town through which it flowed - Rochford, previously called Rocheforte, where the river was forded and therefore the Roach ford provided a new name for the river.

Sea Wall 1 – *stretching for miles along this part of the Essex coast from Burnham all the way round back up to Bradwell*
• *18-50mm lens at 26mm, f16, 1/160 sec, ISO 200, WB5150*

The River Crouch Navigation

Sea Wall 2 – *heading towards the mouth of the estuary. Unlike the south bank of the estuary on which most promontories, bays etc. are named, features on the north bank between Burnham and Holliwell Point remain anonymous.*
• *18-50mm lens at 18mm, f16, 1/30 sec, ISO 100, WB5150*

Estuary Sands – *providing rich pickings for waders and shoreline birds*
• *18-50mm lens at 23mm, f16, 1/50 sec, ISO 100, WB5150*

The River Crouch Navigation

The tidal flats, saltmarsh and coastal grassland and ditch systems also support many species of nationally scarce plant and important populations of rare invertebrates. An important breeding population of grey seal can be found at the mouth of the *Crouch Estuary*.

Consequently the area is subject to a wide range of international, national and regional designations, the *Crouch and Roach Estuaries* are a Site of Special Scientific Interest (SSSI), a Special Protection Area (SPA) and a Ramsar (wetland) site.

Coastal Farming – *where arable farming meets coastal grassland viewed from the sea wall*
• *18-50mm lens at 50mm, f16, 1/50 sec, ISO 100, WB5150*

Sea Wall 3 – *stretching for miles along this part of the Essex coast*
• *70-300mm lens at 92mm, f16, 1/60 sec, ISO 100, WB5150, ND Graduated Filter*

The Essex Coast is one of the top five coastal wetlands in Britain and is internationally important for the wildlife that now abounds, with insects and during the winter months, worms, snails and shellfish all attracting vast numbers of birds.

Feeding across the tidal saltings and sheltering until spring when they return north to their breeding grounds, you may spot flocks of dunlin, lapwing and plovers during winter.

Land that was originally salt marsh was drained and reclaimed from the sea, but in 1897 the sea flooded the farmed fields.

Salt marsh has now returned, leaving only the original parallel lines of the farmer's drainage ditches as evidence.

The seawall and borrowdykes offer fine views across the river and estuary.

Sea Wall 4 – stretching for miles along this part of the Essex coast
• 70-300mm lens at 70mm, f16, 1/50 sec, ISO 100, WB 5150, ND Graduated Filter

Sea Wall 5 – looking back up the river before it reaches the mouth of the estuary • 18-50mm lens at 18mm, f16, 1/15 sec, ISO 100, WB5150, ND Graduated Filter

The sea wall pictured below leads you to Holliwell Point and then on to Bradwell. This part of the coast between the Crouch and the Blackwater is isolated even today. There are many shallow inlets and vast tracts of low marshland and saltmarsh.

This area remained a favourite landing place up until the middle of the 19th century for smugglers.

Now turn back to The Wilderness (page 154), the source of this river: *'from small springs do mighty estuaries grow!'*

The Open Sea – *with Foulness Island leading to Foulness Point in the distance the sea wall sweeps round to Holliwell Point and the North Sea* • 18-50mm lens at 18mm, f16, 1/25 sec, ISO 100, WB5150, ND Graduated Filter

Burnham-on-Crouch

... best known as a yachting centre hosting the internationally famous 'Burnham Week' ...

A historic town situated on the banks of the *River Crouch*.

It has benefited from its location on the coast – first as a ferry port, later as a fishing port known for its oyster beds.

Most recently best known as a yachting centre hosting the internationally famous 'Burnham Week'.

Capital of the picturesque Dengie 100 much of which is unchanged since its listing in the Doomsday Book.

Quayside 1 – a deserted view of one end of the quayside early one morning
• 18-50mm lens at 50mm, f22, 1/30 sec, ISO 100, WB 6000

Houseboats 2 – viewed from a man made shingle jetty early one morning
• 55-200mm lens at 67mm, f22, 1/60 sec, ISO 100, WB 6700, ND Graduated Filter

View of Burnham – viewed from the shoreline near
the marina with the town in the distance
• 10-20mm lens at 10mm, f22, 1/15 sec, ISO 100,
WB 6000, ND Graduated Filter

Quayside 2 – an abstract view of the other end of the
quayside early one morning
• 18-50mm lens at 23mm, f22, 1/25 sec,
ISO 100, ND Graduated Filter

DYNAMIC DIAGONALS

Diagonal lines are a great way of leading the viewer into the picture. When they lead from the edge of the image towards the centre or a significant part of the scene they can help to concentrate the attention onto this area of the image.

Also, you can break the rules and deliberately twist an image to provide a dynamic diagonal and boost the image. Shown left it leads you in bottom left and out again top right.

Thames Estuary

. . . 1.3 mile (2km) long pier which extends out into the Estuary . . .

Southend-on-Sea situated on the north bank of the *Thames Estuary* is the main attraction of all the Essex seaside resorts.

With seven miles of seafront you will find all the traditional seaside delights as well as exciting watersports, theme park thrills, top shopping, thriving arts and culture, glittering night-life and a feast of seafood.

The most notable landmark is that of the 1.3 mile (2km) long pier which extends out into the estuary - it is the longest pleasure pier in the world.

Cliffs Pavillion – one of
Southends entertainment venues
with spring in full swing
• 18-50mm lens at 38mm, f22.0,
1/40 sec, ISO 100, WB 6700

Southend Pier – the world's longest pleasure pier, built in 1830 and stretching about 1.33 miles from shore.
Since 1986, a diesel-hydraulic railway has run the length of the pier, replacing the electric service which opened in 1890.
The pier has been burdened by fires; a fire in 1995 destroyed the bowling alley at the start of the pier and another fire in October 2005 damaged the far end of the pier.
The pier was also run through by a boat in 1984.
• 10-20mm lens at 12mm, f22.0, 1/4 sec, ISO 100, WB 7500

Dinghy Park *– south side of the pier looking into the*
late afternoon sun
* *70-300mm lens at 70mm, f22.0, 1/400 sec, ISO 100,*
  WB 5150, ND Graduated Filter

Westcliffe Hotel *– overlooking*
the well kept Cliff Gardens
where you can gaze out to sea
* *18-50mm lens at 18mm,*
  f16.0, 1/50 sec, ISO 100,
  WB 5150

Thames Estuary

Desolate Beach 1 – *East Beach at low tide*
• 18-55mm lens at 24mm, f22.0,
1/50 sec, ISO 100, WB 6000, ND Graduated Filter

Shoeburyness, situated at the mouth of the busy estuary and east of Southend, boasts of two Blue Flag beaches (see page 31):

East Beach which is a popular sandy/pebbly beach with grass hinterland a quarter of a mile long and

Common Beach (aka West Beach), an ideal location for those wanting to take part in, or watching fun on the waves.

Desolate Beach 2 – *the East Beach at low tide taken early one winter morning at sunrise. This beach is the home of a defence boom, built in 1944, to prevent enemy shipping and submarines from accessing the River Thames. This replaced an earlier similar boom built 100 yards (91 m) to the east. The majority of the boom was dismantled after the war, but around one mile still remains stretching out into the Thames Estuary.*
• 18-50mm lens at 18mm, f22.0, 1/8 sec, ISO 100, WB 5700, ND Graduated Filter

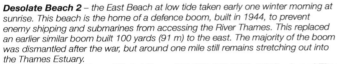

Boat Point – A storage point for rowing boats on East Beach
• 18-55mm lens at 40mm, f22.0,
1/50 sec, ISO 100, WB 5700

SEPIA TONING

There are two methods I use for a sepia effect.

1. To apply straight onto a RAW file I use split toning as described on page 57. I then apply a sepia tint to both the highlights and shadows. There is a balance slider which you can use to favour either the highlights or shadows.

2. If the colour image has already been processed then I use Create New Adjustment Layer pop-up menu and select Hue/Saturation. When the Hue/Saturation dialogue appears, turn on Colourise checkbox, then set Hue to 25 and Saturation to 25 as a starting point. Adjust values to your own taste and Click OK.

Thames Estuary

The Great British Beach Hut is an icon of the great British seaside starting out life as a changing room for modest sea-loving Edwardians.

They go together with ice creams, sandcastles and the unreliable British weather to form part of our experience of summer by the seaside.

Here on this part of the Essex coast they are continuous from Shoeburyness to Thorpe Bay.

Common Beach 1 – based on the Victorian bathing machines, but without the wheels, beach huts were built in their thousands over the 20th century, with their popularity peaking in the 1950s before the sunshine of the 'del sols' lured bathing Britons overseas.
• 18-50mm lens at 50mm, f16.0, 1/60 sec, ISO 100, WB 5150

Common Beach 2 – an ideal location for those wanting to take part in, or watching, fun on the waves
• 18-50mm lens at 29mm, f16.0, 1/40 sec, ISO 100, WB 5150

Common Beach 3 – beach huts have been an icon for generations of holidaymakers. As much a part of our coastal heritage as fish & chips, ice cream and donkey rides
• 18-50mm lens at 23mm, f16.0, 1/60 sec, ISO 100, WB 5150

Common Beach 4 – a few years ago, the beach hut was withering in the depths of unfashionability. Today it has been swept up on a tide of nostalgia and now takes its proper place as an essential part of the British seaside holiday.
• 18-50mm lens at 50mm, f16.0, 1/100 sec, ISO 100, WB 5150

Thames Estuary

For the adventurous photographer **Thorpe Bay** is rich in possibilities with plenty of fishing boats moored in the bay.

The two consecutive days I visited were beautiful, crisp, cold winter mornings and sunrise coincided with a tide that was out but coming in.

The estuary is so flat that the tide comes in very fast and you can find yourself wading back to the shore if you are not careful.

Bay Blues – Thorpe Bay (Day 1) early one January morning. In the winter the low sun rises across the beach nicely side-lighting, or, in these photos backlighting the scene with the sunrise colours reflecting in the wet sands
• 18-50mm lens at 28mm, f18.0, 0.8 sec, ISO 100, WB 5000, ND Graduated Filter

Wintery Bay 1 – Thorpe Bay (Day 2) early one January morning. There was some early morning mist about which warmed up the photos. Lots of small fishing boats are anchored in the bay as well as some of the yachts from the Thorpe Bay yachting club.
• 18-50mm lens at 20mm, f18.0, 1/6 sec, ISO 100, WB 5500, ND Graduated Filter

Wintery Bay 2 – Thorpe Bay (Day 2) early one January morning. The mud here is pretty solid, though wellies are recommended as you sometimes come across softer areas, generously provided by the local bait diggers, to catch out the unwary.
• 18-50mm lens at 26mm, f18.0, 1/10 sec, ISO 100, WB 5150, ND Graduated Filter

Herring Gulls – catching the spray of the waves
• 70-300mm lens at 200mm, f6.3, 1/2000 sec, ISO 400, WB 5150

Gulls, especially chip-stealing herring gulls, are a common sight around Britain's coasts and seaside towns and are increasingly found inland. Evolving more than 15 million years ago in the northern Atlantic and spreading globally. They are typically grey or white with black markings on the head or wings, webbed feet and stout bills.

With around 25 species, the genus contains most of the world's gulls. However, classification has recently become extremely complicated with species being moved between the different gull groups. The great black-backed gull is the largest gull in the world.

Herring Gull – (Larus argentatus) static with a head-on coastal breeze giving an opportunity for a sharp photograph at Southend. Outside the breeding season our commonest coastal large gull which often follows ships and commonly seen at rubbish-tips. Breeds in coastal meadows, dunes, shingle banks, small islands and in some areas on rock ledges and house roofs.
• 70-300mm lens at 300mm, f6.3, 1/2500 sec, ISO 400, WB 5150

Black Headed Gull – (Larus ridibundus) a montage of numerous shots off the coast at Southend. Breeds in colonies at overgrown lake margins, commonly in reed or sedge-beds and on small islands. Widespread and very common at the coast and on inland waters (and fields) during the winter. See page 159 for the gulls in winter plumage
• 70-300mm lens at 248mm, f6.3, 1/4000 sec, ISO 400, WB 5150

Thames Estuary

Old Leigh, with its own unique charm and character, is an old fishing village within Leigh-on-Sea itself.

Situated west of Southend, a picturesque and historical place, having a long established Cockle Industry. Comprising of fishing boats, craft shops, pubs, houses, wharfs, marine industries and cockle sheds which stretch along the shoreline and High Street.

On Reflection 2 – *moored yacht in the creek in Old Leigh at low tide with mirrored reflection*
• *70mm lens, f22.0, 1/15 sec, ISO 100, WB6000*

Early Rising – *sunrise in Old Leigh at low tide with fishing boat moored at a wharf*
• *10-20mm lens at 10mm, f22.0, 1/13 sec, ISO 100, WB 5250*

Fishing Boats – *early morning in Old Leigh at low tide with fishing boats moored on the dock*
• *55-200mm lens at 75mm, f22.0, 1/25 sec, ISO 100, WB 5900*

Moorings at Low Tide – various types of boats moored and
marooned at low tide in the creek
• 18-55mm lens at 46mm, f22.0, 1/20 sec, ISO 100, WB 5500

Two Tree Island – reclaimed saltmarsh, near Old Leigh, leased to Essex Wildlife Trust
and supporting a wide variety of birds, particularly migrants
• 150-500mm lens at 500mm, f8.0, 1/640 sec, ISO 400, WB 4900

During World War II, Leigh-on-Sea was part of a restricted zone and residents who did not have to stay were urged to leave. A balloon barrage was moored to boats in the Thames Estuary to impede efforts by the Luftwaffe to fly up the Thames.

Also, the fishing fleet from Leigh was part of the flotilla from around the country that participated in the evacuation of British forces from the beaches of Dunkirk.

Hadleigh is best known for the Norman, Grade 1 listed, St. James The Less Church in the centre of the town and the ruins of the 13th century **Hadleigh Castle**, painted by Constable in 1829.

Now preserved by English Heritage as a grade 1 listed building. The ruins of the castle provided part of the name for the borough of Castle Point in 1974.

Recent archaeological research has discovered evidence of Roman activity here.

Twin Towers – at sunset one winters evening using a wide angled lens and off-camera flash gun with plenty of patience and experiment
• 10-20mm lens at 11mm, f22.0, ISO 100, WB 6500, ND Graduated Filter, Composite of various shutter speeds

Castle Lights – shortly after sunset with lighting effects. As a starting point set the camera manually to an aperture of f11.0 and a shutter speed of 6 seconds. Check the image and if necessary make adjustments
• 18-55mm lens at 22mm, f11.0, 6 secs, ISO 100, WB 5250

PAINTING WITH FLASH

Set the camera up on a tripod, focus on the foreground using the autofocus and then switch to manual focus. Next select Self-Timer mode, giving you time to fire the shutter and get into painting position (or alternatively use a Remote Controller). Make several exposures using an off-camera flash in different positions on the subject. Because of the long exposures the camera does not pick you up and you will disappear as a blur as you are moving around.

Later, in image editing software, review the RAW images, make any necessary changes, save as tiffs, then layer the images into one file and blend them by selecting 'Lighten' in the Blending Mode.

Hadleigh Castle – the castle formed part of the dower of several English queens in the 15th and 16th centuries, reverting to the king on their deaths. Most notable included Elizabeth Woodville the wife of Edward IV and three of the wives of Henry VIII, Catherine of Aragon, Anne of Cleves, and Catherine Parr.
Edward VI sold the castle in 1551 for the sum of £700 to Lord Rich of Lees priory in Chelmsford who used the castle as a source of stone for other buildings such as churches. The castle later passed from the possession of Lord Rich to the Bernard family.
Years of neglect and the effects of subsidence had left the castle in ruins by the 17th century, but the two towers that were constructed by Edward III still remain to this day.
• 70-300mm lens at 192mm, f22, 1/40 sec, ISO 100, WB 5150

Built in the 1230s, overlooking the *Thames Estuary* and Essex marshes, during the reign of King Henry III for the 1st Earl of Kent and chief justice of England, Hubert de Burgh.

Constructed with Kentish rag stone and cemented by a mortar containing mainly seashells, in particular cockleshells, from the cockle beds of neighbouring Canvey Island.

Substantial additions were made in the mid 14th century by Edward III. It is these later additions that are most visible today which includes the two towers, one almost standing to its original height, the foundations of the great hall, two solars and the kitchen.

Castle Sunset – viewed from one of the towers one spring evening using a wide angled lens and flash gun
• 18-50mm lens at 18mm, f18.0, 0.6 sec, ISO 100, WB 4800, ND Graduated Filter

Below sea level and separated from the mainland by a network of creeks, **Canvey Island** is a reclaimed island in the estuary.

The island was mainly agricultural land until the 20th century when it became the fastest growing seaside resort in Britain between 1911-1951.

Badly flooded in 1953 (costing the lives of 58 islanders) the island is now protected by a concrete sea wall 15 miles long, modelled on Dutch sea defences.

The Sea Wall *– a prominent feature all along the south coast of the island to avert any further flooding tragedies*
• 10-20mm lens at 20mm, f16, 1/30 sec, ISO 100, WB 5150

Hole Haven Jetty *– pointing out towards the end of the pipeline*
• 10-20mm lens at 12mm, f16, 1/40 sec, ISO 100, WB 5150

Canvey is also noted for its relationship to the petrochemical industry. The island was the site of the first delivery in the world of liquefied natural gas by container ship.

Later becoming the subject of an influential assessment on the risks to a population living within the vicinity of petrochemical shipping and storage facilities.

Paddling Pool – *and beach with a view across the estuary to the Kent coast in the background*
• 18-50mm lens at 40mm, f18, 1/30 sec, ISO 100, WB 5150

Hole Haven Pipline – *a UKOP pipeline with one of a few refineries in the background which dominate this part of the Essex coastline*
• 10-20mm lens at 20mm, f16, 1/60 sec, ISO 100, WB 5150

Thames Estuary

Coalhouse Fort, at East Tilbury, is a victorian coastal defence fort overlooking the estuary. Built in 1874 to defend the approaches to London from the threat of invasion from France and other foreign countries. It is said to be one of the finest examples of an armored casemated fort in the United Kingdom.

Since 1983 the fort has been leased to the Coalhouse Fort Project, which is a voluntary organisation, who have worked hard to help keep the fort much of its original architectural form.

Coalhouse Fort 1 – *The fort was built on low lying land in a curve of the River Thames here at East Tilbury and was positioned there to form a 'triangle of fire' between the Essex bank of the river and Cliffe Fort and Shornmead Fort on the Kent bank.*
* *18-50mm lens at 18mm, f16, 1/50 sec, ISO 100, WB 5150*

Coalhouse Fort 2 – in the background with ruins of further coastal defences in the foreground
• 18-50mm lens at 18mm, f16, 1/50 sec, ISO 100, WB 5150

Following the footpath above the shingle of the river, you arrive at what appears to be a water tower, but is in fact a radar tower of World War Two vintage.

Because of the secrecy surrounding radar during World War II this structure was marked on maps as a 'water tower' to confuse the enemy and this 'decoy' name has been continued on maps up until recently.

Nearby you can see the remains of the jetty used when transporting ammunition from Purfleet and also the site of the 1540 blockhouse.

The 'Water Tower' (opposite) *and Jetty* – the disused 'water tower' and abandoned jetty at Coalhouse Point viewed at low tide close to East Tilbury Marshes with the Thames flowing towards London and the Kent coast in the distance
• 18-50mm lens at 18mm, f16, 1/60 sec, ISO 100, WB 5150

The Shoreline at East Tilbury Marshes 1 – *viewed looking back to the 'Water Tower' with more coastal defences in the foreground. Some time before the threat of invasion officially ceased, wartime defensive structures began to disappear. Trenches were filled and concrete tank traps removed, however, getting rid of the pillboxes proved far harder. Coastal pillboxes have fallen into the sea as the elements have eaten away the coastline.*
• *18-50mm lens at 18mm, f16, 1/80 sec, ISO 100, WB 5150*

After passing by the power station we come to the present **Tilbury Fort** which was built in 1672 during the reign of Charles II, and is the most impressive example of a seventeenth century bastioned artillery fort and military engineering remaining in Britain.

The fort is a massive complex of earth banks, walls, moats and tunnels, most of which survive intact. Never being used in battle, the fort did its job simply through its intimidating presence.

Now preserved by English Heritage giving everyone an opportunity to explore the history of the defence of the Thames.

Tilbury Power Station 1 – *approaching the power station with the Thames at low tide and the marshes to the right*
• *18-50mm lens at 23mm, f16, 1/80 sec, ISO 100, WB 5150*

The Shoreline at East Tilbury Marshes 2 – viewed looking towards Tilbury Power Station with a disused jetty in the foreground and the Kent coast in the background
• 18-50mm lens at 18mm, f16, 1/60 sec, ISO 100, WB 5150

Tilbury Power Station 2 – photographed from the Thames viewing point outside the Water Gate entrance to Tilbury Fort. Of course, a visual illusion, the power station is situated a few hundred yards to the east of the fort and the gun enplacement
• 55-200mm lens at 120mm, f18, 1/50 sec, ISO 100, WB 5500

Thames Estuary

Tilbury Jetty – used for large vessels to unload at the power station photographed from the Thames viewing point outside the Water Gate entrance to Tilbury Fort
• 55-200mm lens at 130mm, f18, 1/80 sec, ISO 100, WB 5150

Tilbury Fort 1 – gun enplacement on the East Gun Line looking out over the Thames with Kent in the background
• 10-20mm lens at 10mm, f18, 1/125 sec, ISO 100, WB 5150

Water Gate – architectural detail
• 70-300mm lens at 192mm, f18, 1/60 sec, ISO 100, WB 5150

Henry VIII first built a fort here, and his daughter Elizabeth I gave her famous rousing speech to English troops near Tilbury Fort in 1588 as the Spanish Armada approached along the Channel. No doubt the backdrop of Tilbury lent additional power to Elizabeth's words:

'I am come amongst you, as you see, at this time, not for my recreation and disport, but being resolved, in the midst and heat of the battle, to live and die amongst you all; to lay down for my God, and for my kingdom, and my people, my honour and my blood, even in the dust. I know I have the body but of a weak and feeble woman; but I have the heart and stomach of a king, and of a king of England too...'

THE END

Tilbury Fort 2 – gun enplacement on the East Gun Line with Kent in the background and the Thames making its way into London
• 18-50mm lens at 20mm, f18.0, 1/60 sec, ISO 100, WB 5150

Bradwell Power Station – a view across the Blackwater Estuary from Mersea Island with the now de-commissioned power station at Bradwell on the skyline
• 10-20mm lens at 10mm, f16.0, 1/15 sec, ISO 100, WB 5150, ND Graduated Filter

Beaumont Quay *– a view across Beaumont Cut on Hamford Water from the quayside at low tide*
• 18-50mm lens at 24mm, f16.0, 1/125 sec, ISO 100, WB 5150, ND Graduated Filter

Acknowledgements

Many thanks to Steve Hedges of Artescape Photography (artescapephotography.com) who has been a good friend and mentor, whose enthusiasm for photography has certainly rubbed off.

My friends within the 20/20 Exhibition Group whose encouragement has been appreciated.

All those who kindly gave me permission to use material for this book.

My family and friends who have had to endure me talking about the locations etc. and for giving me their honest critiques.

CPSIA information can be obtained
at www.ICGtesting.com
Printed in the USA
LVIW020011070612

2871LVUK00005BB